AI DENTAL MARKETING

How AI Is Disrupting Digital Marketing, Search, and Patient Acquisition

A systems-level guide for dental practice owners who want clarity—not tactics, hype, or shortcuts

Why visibility works the way it does

Why some efforts compound while others stall

Why alignment matters more than activity

And how AI amplifies structure—for better or worse

By Chase Powell

This book explains how digital marketing systems actually work—

so you can make informed decisions about vendors, budgets, timelines,

and the role AI should (and should not) play in your practice.

AI Dental Marketing

ISBN: 978-1-960410-22-1

Printed in the United States of America

First Edition

Author Website: https://patientflow.us/book-dental

Table of Contents

Author's Note

This book exists because too many smart dental practice owners feel confused about something that shouldn't be mysterious.

Not because marketing is easy.

But because it's been made noisy.

Over the last several years—especially with the rise of AI—digital marketing has been framed as either magic or manipulation. Either you're told there's a secret formula you don't have, or you're warned that everything is changing so fast that understanding it is pointless.

Neither is true.

What *is* true is that the systems governing visibility, trust, and patient acquisition have become more powerful—and less forgiving—at the same time. They reward clarity. They punish contradiction. And they amplify whatever structure already exists inside a business, for better or worse.

That's why this book doesn't try to teach you how to "do marketing."

It tries to help you **see it clearly enough to make good decisions**.

Why this is not a tactics book

You won't find:

- step-by-step SEO checklists
- ad platform walkthroughs
- content calendars
- tool recommendations disguised as advice

Not because those things don't matter—but because they only matter *after* you understand the system they operate inside.

Too many owners have been handed tactics without context. They've been told what to do without being told *why* it might work, *when* it won't, or *what* it depends on. The result is frustration, vendor churn, and the nagging sense that marketing is harder than it should be.

This book is meant to remove that fog.

Why dental practices are different (and why that matters)

Dentists sit in a rare and sensitive category.

Patients aren't just buying a product.

They're making personal, visible, and often emotional decisions.

That reality changes how platforms behave.

Search engines, ad systems, social platforms, and now AI-driven discovery tools all treat trust-sensitive categories differently.

They are more cautious. They rely more heavily on corroboration. And they respond more strongly to consistency—or the lack of it.

Understanding that isn't about gaming the system.

It's about **aligning with how the system is designed to reduce risk**.

What AI actually changes—and what it doesn't

AI didn't invent these dynamics.

It accelerated them.

AI makes it easier to:

- analyze patterns
- compare signals
- surface contradictions
- scale whatever already exists

Which means it rewards businesses that already make sense—and exposes those that don't.

This is why AI feels destabilizing to some owners and empowering to others. The technology isn't choosing winners. It's **amplifying structure**.

That's why this book treats AI as a force multiplier, not a shortcut.

How to read this book

You don't need to read this cover to cover in one sitting.

In fact, most people won't.

Some chapters are meant to be reference points.

Some are meant to shift how you think.

Some are meant to help you evaluate advice you're already receiving.

If at any point you find yourself thinking:

> "This is simpler than I thought—but I understand why it's hard,"

then the book is doing its job.

What I hope you take away

By the end, you should:

- understand how visibility actually works
- see why some efforts compound and others stall
- feel more confident evaluating vendors and strategies
- recognize that alignment matters more than activity
- view AI as something to integrate carefully, not chase blindly

Most importantly, I hope marketing feels **legible**.

Not easy.

Not guaranteed.

But understandable.

When you can see the system clearly, you don't need certainty to move forward. You need judgment.

That's what this book is here to support.

If it helps you make calmer, clearer, more intentional decisions—then it's done what it was meant to do.

About the Author

Chase Powell is a digital marketing strategist who helps dental practice owners make sense of a marketing landscape that has become louder, faster, and increasingly shaped by AI-driven systems.

Rather than focusing on tactics or short-term performance, Chase's work centers on understanding how modern platforms evaluate trust, reduce risk, and decide which businesses get recommended. His perspective has been shaped by years spent inside search, local visibility, paid media, and content systems—watching what compounds quietly over time, and what creates the illusion of progress without durability.

Chase believes most marketing frustration doesn't come from lack of effort or intelligence, but from misunderstanding incentives. Search engines, Maps, ad platforms, and AI tools are not designed to help businesses grow; they are designed to protect users from regret. When owners understand that distinction, marketing decisions become calmer, clearer, and far less reactive.

He is the founder of **Patient Flow**, where he develops clarity-first frameworks and diagnostic tools for dental practice owners who want to understand their marketing systems before investing more money, time, or trust in execution. His approach avoids hacks, guarantees, and automation-for-automation's-sake, especially in regulated, trust-sensitive categories like dental procedures.

AI Dental Practice Marketing reflects Chase's core philosophy: that marketing works best when it is understood as a system—one that rewards consistency, coherence, and credibility more than speed, volume, or novelty.

This book was written for owners who don't want more tactics, but better judgment.

CHAPTER 1

The New Reality for Dental Practice Patient Acquisition

Why this chapter matters

If you run a dental practice today, there's a good chance marketing feels harder than it used to — even if you're doing more than ever.

You might not be able to point to a single breaking point. There was no obvious collapse. No dramatic penalty. No moment where everything stopped working at once.

Instead, it feels like this:

- What used to be reliable feels unpredictable
- What used to scale feels fragile
- Advice feels louder, more confident — and less useful
- Everyone is selling urgency, but no one is explaining *why*

This chapter exists to name what's actually happening — without panic, hype, or false certainty.

Because the problem isn't that dental practice marketing is broken.

The problem is that **the ground rules changed**, and most owners were never given a clear explanation of how or why.

Why everything suddenly feels unstable

From the outside, it looks like chaos:

- Search results change shape
- Maps behave inconsistently
- Ads cost more and convert less predictably
- AI shows up in places it didn't exist before

But from inside the system, something much simpler is happening.

Digital marketing has entered a phase where:

- **Speed increased**
- **Signal scrutiny increased**
- **Tolerance for confusion decreased**

None of those are random.

They're the natural result of platforms trying to do one thing better than before:

reduce user regret at scale.

When that goal tightens, everything downstream feels less forgiving.

AI didn't cause this — it exposed it

It's tempting to blame AI for the instability.

After all:

- AI-generated answers are appearing in search
- Content is easier to produce than ever
- Tools promise automation, scale, and shortcuts

But AI didn't create the underlying pressure.

AI **accelerated** it.

Long before AI:

- SEO rewarded consistency over time
- Maps favored established patterns
- Ads borrowed trust from organic presence
- Fragmented strategies underperformed aligned ones

AI simply made those truths harder to ignore — and faster to enforce.

What used to take years to reveal now shows up in months.

What used to be masked by effort is now surfaced by pattern recognition.

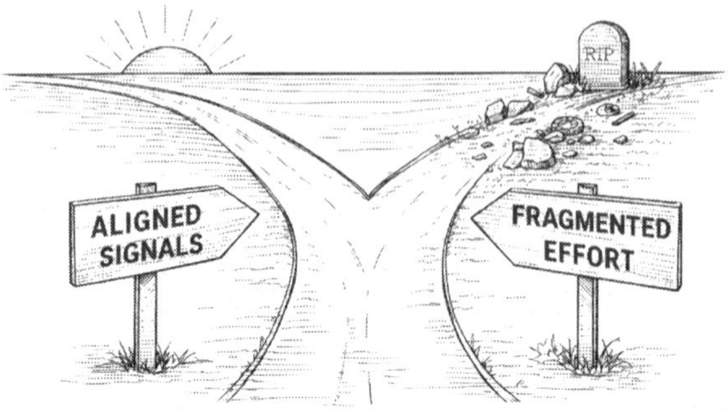

Why this moment feels different for dental practices

Every industry experiences platform shifts.

But dental practices feel this one more acutely for a few reasons:

- Decisions are personal and visible
- Trust matters more than novelty
- Outcomes carry emotional weight
- Local context is unavoidable

In other words, dental practices sit in a **high-trust, moderate-risk category** — exactly the kind of category platforms become more conservative about as systems mature.

That doesn't mean opportunity disappears.

It means **clarity becomes the price of admission**.

The real cost isn't doing nothing — it's doing the wrong thing

Most owners don't stall because they refuse to act.

They stall because they act without a clear mental model.

That leads to:

- chasing new channels without fixing foundations
- switching vendors without changing structure
- adding tools instead of resolving contradictions
- mistaking activity for progress

Over time, this creates fatigue — not failure, but drift.

The danger isn't standing still.

It's moving fast **in conflicting directions**.

Why "what worked before" stopped being a strategy

This is one of the hardest truths to accept — especially for successful operators.

If something worked for years, it *should* still work... right?

Not necessarily.

Digital marketing systems don't reward history.

They reward **current alignment**.

Past success helps — but only if:

- signals are still consistent
- services are still clearly defined
- messaging hasn't drifted

- platforms can still infer the same story

When those things slip — even subtly — momentum slows.

Not as punishment.

As caution.

The real question isn't success or failure — until it isn't

At first glance, the title sounds dramatic.

But the real divide isn't luck.

It isn't budget.

It isn't even execution quality.

It's **understanding**.

Owners who understand how modern digital systems infer trust:

- make calmer decisions
- invest more deliberately
- avoid chasing noise
- use AI as leverage, not a crutch

Owners who don't:

- feel pressured to move faster
- overreact to volatility
- outsource judgment
- accumulate disconnected efforts

The outcomes diverge quietly — then suddenly.

Decision-level insight

This moment is not a collapse.

It's a **filter**.

AI and modern platforms are not asking:

> "Who is trying hardest?"

They're asking:

> "Who makes the most sense?"

That distinction determines whether this era feels like a gold rush or a graveyard.

The next chapter explains **why dental practices, specifically**, are positioned on the favorable side of this shift — and why understanding your category's structural advantages matters more now than ever.

CHAPTER 2

Why Dental Practices (And Why Now): The Category Mechanics Behind the Attention

Why this chapter matters

After reading the first chapter, a reasonable question lingers:

> *"If everything feels harder, why would dental practices be well positioned at all?"*

This chapter answers that directly.

Because while marketing systems are becoming stricter, faster, and less tolerant of confusion, **they are not neutral across categories**. Some types of businesses align naturally with how modern platforms evaluate trust, intent, and risk. Others fight those incentives constantly.

Dental practices, structurally, sit in a surprisingly strong position —*if* owners understand what the system values and why.

This chapter isn't about optimism.

It's about mechanics.

Platforms don't evaluate businesses equally

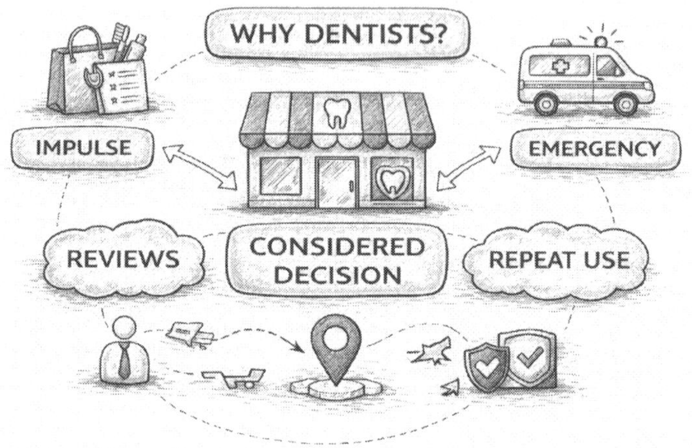

Search engines, ad systems, and AI-driven discovery tools don't just evaluate *quality*. They evaluate **risk**.

Every category is silently scored along dimensions like:

- consequence of a bad recommendation
- likelihood of repeat interaction
- clarity of intent
- local relevance
- trust sensitivity

A wrong restaurant recommendation is inconvenient.

A wrong recommendation feels personal.

That difference changes everything about how platforms behave.

Dentists are not punished for this.

They're **filtered more carefully**.

Dental practices live in the "considered decision" zone

Dental practices sit in a specific behavioral category that platforms understand well:

- Not impulse-driven
- Can be emergency-driven
- Not purely educational
- Not purely transactional

Instead, dental practice decisions are:

- local
- repeatable
- trust-based
- consideration-heavy
- frequently motivated by pain or deterioration

People don't choose a dental practice the way they choose a café.

But they also don't choose one the way they choose a surgeon.

That middle ground matters.

It creates:

- high-intent searches
- meaningful comparison behavior
- reliance on Maps, reviews, and corroboration
- opportunities for trust to compound over time

Platforms *like* this kind of demand—because it's predictable, legible, and defensible.

Repeat behavior changes how systems allocate trust

One of the quiet advantages of dental practices is **repeat demand**.

Patients don't usually interact once and disappear forever. They:

- return for maintenance
- explore adjacent services
- build familiarity with a provider

From a systems perspective, repeat behavior is powerful.

It suggests:

- satisfaction
- legitimacy
- reduced regret

- stable outcomes

Businesses with repeat interaction patterns are safer to surface than one-off vendors chasing novelty.

This is one reason dental practices respond so strongly to alignment over time:

the system *expects* continuity.

Local intent is not a limitation—it's leverage

Many owners see "local" as a constraint.

In reality, locality is one of the strongest trust signals platforms have.

Local intent allows systems to:

- narrow choices
- validate physical presence
- use proximity as a confidence filter
- observe real-world behavior

Dental practices benefit from this because:

- proximity matters to patients
- travel tolerance is moderate
- legitimacy is tied to place, not just branding

This gives Maps and local signals more weight—and makes holistic alignment even more important.

Why investors noticed dental practices early

This isn't accidental.

Investors tend to look for categories with:

- predictable demand
- repeat usage
- strong margins
- defensible local positioning
- operational leverage

Dental practices check those boxes.

But what's often missed is *why* that matters for marketing.

Categories that scale operationally tend to scale **digitally**—*if* their trust signals are coherent.

The same characteristics that make dental practices attractive to investors make them legible to modern discovery systems.

Why AI systems "like" dental practice-style businesses

AI systems don't favor industries.

They favor **patterns**.

Dental practices generate patterns that are easier to learn from:

- consistent service categories
- standardized terminology

- recurring patient questions
- repeat interactions
- stable local entities

That makes it easier for AI to:

- summarize accurately
- infer expertise
- identify legitimate providers
- avoid risky recommendations

In other words, dental practices are **structurally compatible** with AI-driven discovery—*when their signals agree.*

The opportunity is real—but conditional

This is where false optimism can creep in, so it's important to be precise.

Dentists are well positioned **only if**:

- their services are clearly defined
- their messaging is consistent
- their online presence corroborates itself
- their operations support trust

When those conditions are met, modern systems amplify them efficiently.

When they aren't, the same systems become cautious—and visibility stalls.

The opportunity exists.

But it is not automatic.

Why this moment favors understanding over aggression

In earlier eras of digital marketing, aggressive tactics could outpace understanding.

You could:

- outspend competitors
- outproduce content
- exploit temporary gaps

Those strategies are less reliable now.

As systems mature, they increasingly reward:

- coherence
- restraint
- consistency
- clarity over time

Dentists that understand this don't need to race.

They need to **align**.

Decision-level insight

Dental practices are not late to the game.

They're early to a phase where:

- trust matters more than volume
- systems reward consistency over creativity
- AI amplifies clarity faster than confusion

This is not a gold rush for anyone willing to spend more or post more.

It's a gold rush for owners who understand **why their category works**—and build marketing systems that agree with that reality.

The next chapter clears away the biggest obstacle to taking advantage of this position: **marketing myths that persist precisely because they sound plausible**, even when they conflict with how the system actually behaves.

CHAPTER 3

The Myth Cleanup: Why "Channel-Based Marketing" Persists

Why this chapter matters

By the time most dental practice owners start questioning their marketing, they're not confused about effort.

They're confused about *structure*.

They've done SEO.

They've run ads.

They've posted on social.

They've hired vendors who "specialize."

And yet, nothing quite adds up.

This chapter exists to explain **why that confusion is so common**, why it isn't a personal failure, and why one of the most persistent ideas in digital marketing—*channel-based thinking*—quietly breaks results over time.

The appeal of channels (and why they're misleading)

Channel-based marketing feels logical.

SEO lives here.

Ads live there.

Social is over here.

Email is separate.

Each channel:

- has its own metrics
- its own tools
- its own experts
- its own reports

This separation makes marketing feel manageable.

It also hides the most important truth:

> **Platforms do not experience your marketing in channels.**
>
> **They experience it as a single pattern.**

Channel thinking is a convenience for humans.

It is not how systems infer trust.

How channel silos were accidentally created

This isn't because marketers are incompetent.

It's because of history.

Digital marketing evolved in layers:

- search first
- then ads
- then social
- then mobile
- now AI

Each layer created new tools, new specialists, and new economic incentives.

Over time:

- agencies specialized
- platforms built isolated dashboards
- reporting reinforced separation

No one designed silos intentionally.

They **emerged naturally**.

Unfortunately, systems evolved past them.

Why vendors benefit from channel separation

This is uncomfortable—but important.

Channel separation makes selling easier.

It allows vendors to say:

- "SEO isn't working, but ads will"
- "Ads are expensive, but social will fix it"
- "Social doesn't convert, but content will"
- "Content doesn't rank, but backlinks will"

Each promise sounds plausible *in isolation*.

What gets lost is the question:

> "Do these efforts agree with each other?"

Fragmentation creates motion.

Alignment creates momentum.

Most vendors are incentivized to deliver activity inside their lane —not coherence across the system.

Why dashboards reinforce the illusion

Platform dashboards are persuasive.

They show:

- clicks
- impressions

- conversions
- graphs moving upward

What they rarely show is **interaction between channels**.

A dashboard can't easily answer:

- Did social validate organic trust?
- Did content reduce ad friction?
- Did Maps legitimacy improve conversion quality?

So owners see:

- "SEO is up"
- "Ads are down"
- "Social engagement is flat"

And assume something is broken.

In reality, the system may simply be receiving **mixed signals**.

Why channel success doesn't equal system success

This is where frustration peaks.

A dental practice might have:

- a technically sound website
- well-run ads
- consistent posting
- decent reviews

And still feel stuck.

Because success inside a channel does not guarantee:

- clarity across channels
- consistency of message
- reinforcement of trust

Platforms don't reward isolated excellence.

They reward **agreement**.

The quiet cost of contradiction

Contradictions are rarely obvious.

They show up as:

- slightly different service emphasis
- inconsistent language
- mismatched visuals
- promises that don't fully align

No single contradiction sinks a business.

But systems don't average contradictions.

They **hesitate because of them**.

And hesitation looks like:

- stalled rankings
- volatile Maps visibility

- rising ad costs
- slower growth

Nothing is "wrong."

The system is just unsure.

Why owners get blamed unfairly

When results stall, blame tends to flow downhill.

Owners are told:

- "You need more content"
- "You need more budget"
- "You need to be more consistent"
- "You need to trust the process"

Those statements may be true.

But they often miss the root issue.

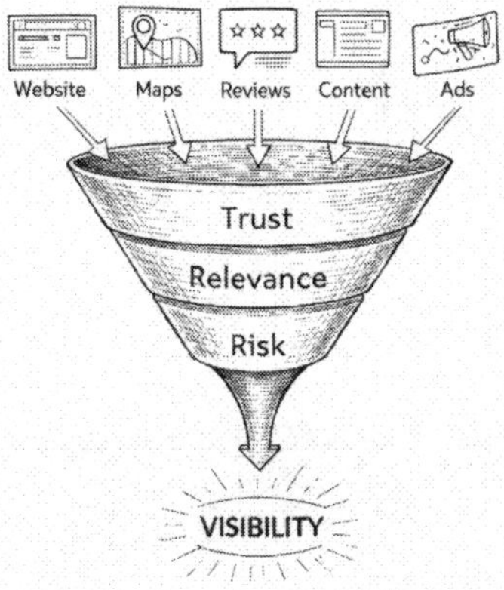

More effort inside a fragmented system doesn't fix fragmentation.

It **amplifies it**.

That's why this book insists on understanding *systems before tactics*.

Introducing signal alignment (without buzzwords)

Here's the core idea this chapter introduces—one that the rest of the book builds on:

> Marketing works when **signals across channels tell the same story**.

Not louder.

Not more often.

Not everywhere.

Just *the same*.

Maps supporting the website.

Website clarifying content.

Content reinforcing social.

Social validating reality.

Ads accelerating what already makes sense.

This isn't a strategy.

It's a condition.

When it exists, marketing compounds.

When it doesn't, marketing feels busy and brittle.

Why this myth survives even when it fails

Channel-based thinking persists because:

- it's teachable
- it's sellable
- it produces activity
- it feels concrete

Holistic thinking is harder:

- it requires judgment
- it resists checklists
- it forces prioritization

But systems don't reward what's easy to explain.

They reward what's **easy to understand at scale**.

And coherence is easier to understand than complexity.

Decision-level insight

You don't have a "SEO problem" or an "ads problem" or a "social problem."

You have either:

- a system that agrees with itself
- or one that doesn’t

Once you stop thinking in channels, the rest of the book clicks into place.

If you want to see your marketing as a system rather than a collection of channels, you can start with a simple system review. Some readers choose to use a structured self-assessment—such as the system clarity tools at **https://patientflow.us/resources-dental** —to make contradictions visible before adding more effort.

The next chapter explains **what the system actually shows users when they search**, and why understanding the structure of the SERP is the foundation for every decision that follows.

CHAPTER 4

What a SERP Really Is (And Why It's Not "Google")

Why this chapter matters

Most marketing decisions fail *before* tactics ever come into play—because they're made with the wrong mental model of how search actually works.

When dental practice owners say things like:

- "We're not showing up on Google," or
- "Our SEO company says we're ranking,"

they're usually talking past the system without realizing it.

This chapter exists to give you a **shared map**—a clear, non-technical understanding of what actually happens when someone searches, where attention goes, and why visibility is not a single thing. Without this map, every conversation about SEO, Maps, ads, or AI becomes confusing, contradictory, or easy for someone else to control.

Before we talk about *how* to compete, you need to understand **where competition actually happens**.

The SERP is not "Google"

A SERP is a **Search Engine Results Page**.

That sounds obvious—but the misunderstanding runs deep.

When someone searches for something like *"Invisalign near me"*, they're not seeing "Google" in a general sense. They're seeing a **designed results environment**—a page intentionally structured to answer a question *as efficiently as possible* while managing user risk, intent, and trust.

Think of the SERP less like a list and more like a **control panel**.

It's made up of multiple sections, each with a different job:

- Some sections are designed to *compare*
- Some to *validate trust*
- Some to *resolve location*
- Some to *sell immediacy*

What matters is not whether you "show up," but **where you show up and why**.

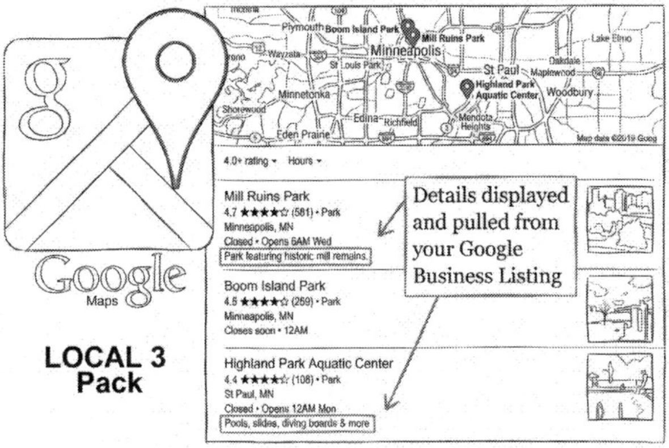

Google
Maps
LOCAL 3
Pack
Minneapolis
St Paul
Minnetonka
Eden Prairie
Edina
Woodbury
4.0+ rating
Hours
Mill Ruins Park
4.7 ★★★★☆ (581) · Park
Minneapolis, MN
Closed · Opens 6AM Wed
Park featuring historic mill remains.
Boom Island Park
4.5 ★★★★☆ (259) · Park
Minneapolis, MN
Closes soon · 12AM
Highland Park Aquatic Center
4.4 ★★★★☆ (108) · Park
St Paul, MN
Closed · Opens 12AM Mon
Pools, slides, diving boards & more
Details displayed and pulled from your Google Business Listing

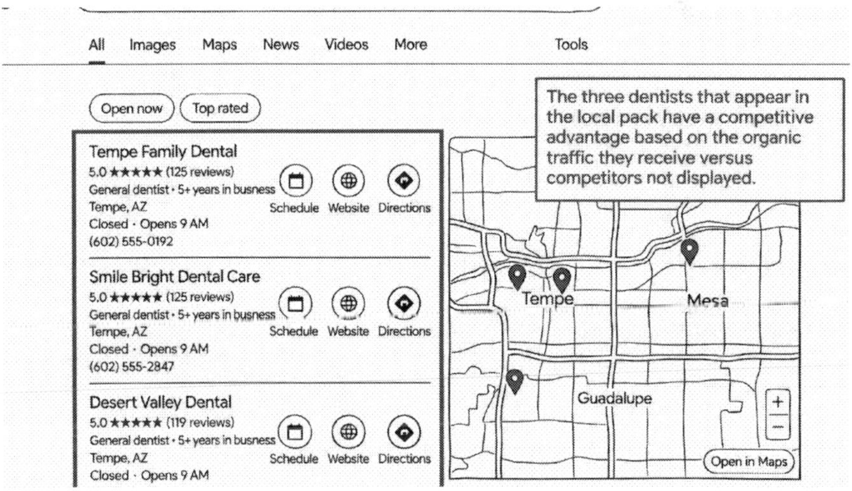

All
Images
Maps
News
Videos
More
Tools
Open now
Top rated
Tempe Family Dental
5.0 ★★★★★ (125 reviews)
General dentist · 5+ years in busness
Tempe, AZ
Closed · Opens 9 AM
(602) 555-0192
Schedule
Website
Directions
Smile Bright Dental Care
5.0 ★★★★★ (125 reviews)
General dentist · 5+ years in busness
Tempe, AZ
Closed · Opens 9 AM
(602) 555-2847
Schedule
Website
Directions
Desert Valley Dental
5.0 ★★★★★ (119 reviews)
General dentist · 5+ years in busness
Tempe, AZ
Closed · Opens 9 AM
Schedule
Website
Directions
The three dentists that appear in the local pack have a competitive advantage based on the organic traffic they receive versus competitors not displayed.
Tempe
Mesa
Guadalupe
Open in Maps

The four primary SERP surfaces (for dental practices)

While SERPs can vary, most high-intent local searches relevant to dental practices break into four core areas:

1. Paid Ads (Top and Bottom)

These are the sponsored placements—clearly labeled, auction-based, and immediate.

Their job is **speed**.

Not trust. Not completeness. Speed.

Ads exist because some searchers want to act *now*, and some businesses are willing to pay to be present at that moment.

Important distinction:

Ads don't replace organic trust signals. They **sit on top of them**.

2. The Local Pack (Maps)

This is the map-based result showing a small number of nearby businesses, typically with:

- Reviews
- Distance
- Categories
- Quick actions (call, directions)

For dental practices, this is often the **most valuable real estate on the page**—and also the most misunderstood.

Maps results are not "mini organic listings."

They are governed by **local relevance, proximity, and behavioral signals**, not just website strength.

We'll unpack this fully in the Maps chapter, but for now, understand this:

The local pack is its own ecosystem.

3. Organic Results (Traditional SEO)

These are the familiar blue links (and increasingly rich results) that live *below* ads and maps.

Their job is **depth and explanation**:

- Services
- Conditions
- Comparisons
- Long-form answers

Organic rankings are where trust is *demonstrated over time*, not bought.

4. Secondary Surfaces (FAQs, AI summaries, videos, images)

These include:

- "People also ask"
- AI-generated summaries
- Videos

- Images
- Expanded answers

These surfaces are growing—not because Google wants to replace businesses, but because many queries are better served with **synthesized information**.

This matters more in educational searches than transactional ones—but the boundary is blurring.

Attention is not evenly distributed (and never has been)

One of the most persistent myths in marketing is that "being on page one" is the goal.

It isn't.

Attention follows a steep curve:

- A small number of results receive the majority of clicks
- The rest receive dramatically less
- Page two is functionally invisible for high-intent searches

This isn't a failure of SEO.

It's a feature of human behavior.

People searching for dental practice services are not browsing for fun. They are:

- Evaluating risk

- Looking for reassurance
- Seeking proximity and legitimacy
- Provider offerings

The SERP is designed to support *those instincts*, not fairness.

Why page two doesn't matter (for dental practices especially)

In low-risk searches, people may explore.

In **consideration-based healthcare-adjacent searches**, they don't.

When someone is choosing a provider for an invasive dental treatment::

- Fewer options feel safer
- Familiar signals feel trustworthy
- Distance feels meaningful

This is why:

- "Almost ranking" produces almost no value
- Incremental ranking gains can suddenly matter *a lot*
- Visibility is discontinuous, not gradual

You don't slowly win traffic.

You cross thresholds.

Personalization: why your own searches lie to you

One of the most dangerous habits in marketing decision-making is this sentence:

> "When I search, I see us."

You are not the market.

Search results are influenced by:

- Your location
- Your search history
- Your past interactions
- Your device

This is especially true in Maps.

So when owners or staff use personal searches to evaluate performance, they are often seeing a **customized reality**, not a neutral one.

The system is not lying—but it is not showing you what *everyone else* sees either.

The most important reframing

Here's the core takeaway of this chapter:

> A SERP is not a scoreboard.
>
> It is a decision environment.

Every element on the page exists because it helps the platform:

- Reduce user uncertainty
- Resolve intent quickly
- Maintain trust

Your marketing efforts are not rewarded for effort.

They're rewarded for **alignment**.

Once you see the SERP this way, a lot of confusion disappears:

- Why great websites sometimes don't rank
- Why ads feel powerful but fragile
- Why Maps behaves "strangely"
- Why AI summaries appear for some searches and not others

This chapter gives you the map.

The next chapters explain **what powers the engine** behind it—starting with intent, and why what people *mean* matters more than what they type.

Decision-level insight

If you don't know **which surface you're competing on**, you can't evaluate:

- SEO progress
- Ad performance
- Vendor claims

- AI impact

Understanding the SERP doesn't make marketing easy—but it makes it *legible*.

And legibility is the foundation of good decisions.

Great. Then we'll intentionally **shift gears here**—less diagramming, more reasoning out loud—while still keeping the book disciplined and clear.

CHAPTER 5

Intent Is the Engine: What People Mean When They Search

Why this chapter matters

If Chapter 4 gave you the *map*, this chapter explains the force that moves everything on it.

Most marketing conversations get stuck on *what people type*:

- keywords
- phrases
- rankings

But platforms don't actually optimize around words.

They optimize around **intent**.

That difference—between what someone types and what they *mean*—is where most dental practice marketing either starts working or quietly breaks down. If you misunderstand intent, everything downstream looks confusing: rankings feel unstable, ads feel expensive, and traffic doesn't behave the way you expect.

This chapter exists to make intent feel obvious—because once it does, a lot of platform behavior stops feeling arbitrary.

Searches are questions, not strings of words

When someone types something into Google, they're not offering a keyword. They're asking a question.

Sometimes that question is explicit:

- "Does a root canal hurt?"
- "How long do dental implants last?"

Sometimes it's compressed:

- "Dentist near me"
- "Teeth whitening cost"

And sometimes it's barely articulated at all:

- "Dentist"
- "Dental clinic"

The system's job is not to match words.

It's to **infer the problem behind the words** and reduce uncertainty as quickly as possible.

That's why two searches that look similar on paper can behave very differently in practice.

High intent vs. low intent (in real dental practice terms)

Intent is often explained academically—*informational, commercial, transactional*. That's technically correct, but not very useful.

A more practical way to think about it for dental practices is **risk and readiness**.

Some searches signal:

- curiosity
- early research
- low commitment

Others signal:

- urgency
- comparison
- willingness to act

For example:

Someone searching "What is a root canal" / "Emergency dentist near me" is not choosing a provider.

They're deciding what belongs in their mental world at all.

Someone searching *"Invisalign near me"* has already crossed that line.

They're now choosing **who to trust**.

Those two searches may share a word, but they do not share intent—and the system treats them very differently.

Why intent controls where you appear on the SERP

This is where Chapter 4 and Chapter 5 lock together.

Different SERP surfaces exist to satisfy **different kinds of intent**:

- Maps help people resolve *location and legitimacy*
- Organic listings help people *understand and compare*
- Ads help people *act quickly*

The system isn't asking:

> "Which business deserves to rank?"

It's asking:

> "What would help this person feel confident enough to move forward?"

That's why:

- Some searches trigger a heavy local pack
- Some trigger long-form organic content
- Some surface AI summaries
- Some show ads aggressively—and others barely at all

Intent determines the mix.

Branded vs. non-branded searches are really about trust

One of the most misunderstood distinctions in marketing is branded vs. non-branded search.

It's often framed as:

- "Branded traffic is good"
- "Non-branded traffic is growth"

That's not wrong—but it misses the deeper point.

Branded searches exist because **trust already happened somewhere else**.

When someone searches for a specific dental practice by name, they're not evaluating options. They're confirming details:

- hours
- location
- phone number

Non-branded searches are where **trust is being negotiated**.

That's where:

- reviews matter more
- proximity matters more
- authority matters more

This is also where the system becomes more conservative—because the risk of sending someone to the *wrong* provider is higher.

Why "educational content" is not a growth strategy by itself

This is where a lot of well-intentioned marketing goes sideways.

Educational content feels safe. It feels helpful. It feels aligned with expertise.

But education alone does not equal acquisition.

Most educational searches sit **upstream** of decision-making. They shape awareness, not choice. That doesn't make them useless—but it does mean they play a different role.

When owners hear:

> "Content builds trust"

They often assume:

> "Any content builds patients"

The system doesn't work that way.

Content builds trust **when it aligns with intent**—not when it exists in isolation.

Why platforms care so much about intent (and not your goals)

This is an important recalibration.

Search platforms are not trying to:

- help your business grow
- reward effort
- be fair

They are trying to:

- minimize user regret
- reduce bad outcomes
- keep people coming back

In dental practice searches, that means:

- being cautious with recommendations
- preferring established signals
- favoring clarity over novelty

Intent helps platforms decide how much risk they can tolerate on the user's behalf.

The higher the perceived risk, the more conservative the system becomes.

The quiet mistake that creates most frustration

Here's the pattern that causes the most confusion:

A business optimizes heavily for **low-intent visibility**, then expects **high-intent results**.

That shows up as:

- traffic without bookings

- rankings without calls
- impressions without outcomes

Nothing is "broken."

The system is doing exactly what it's designed to do.

It's just responding to a different question than the one the business thinks it's answering.

Decision-level insight

Intent is not a keyword property.

It's a **human state** the system is trying to interpret.

Once you see that:

- SERP behavior makes sense
- Channel differences make sense
- Why some efforts compound and others stall becomes clearer

You don't win visibility by talking louder.

You win it by **aligning with what people are actually trying to decide**.

The next chapter builds on this by showing how that intent concentrates demand into a surprisingly small number of searches—and why most marketing effort misses them entirely.

CHAPTER 6

Where Traffic Actually Comes From (And Why "More Keywords" Fails)

Why this chapter matters

If intent is the engine, this chapter explains where the fuel actually enters the system.

Most dental practice marketing frustration comes from a quiet mismatch between effort and opportunity. Owners are often doing *a lot*—creating content, running ads, publishing pages—yet the results feel thin, inconsistent, or disconnected from growth.

This chapter exists to explain why.

Not because platforms are unfair.

Not because SEO is broken.

But because **demand is far more concentrated than most people realize**—and most marketing effort is pointed at the wrong part of the curve.

Once you see how traffic really distributes, a lot of common advice stops making sense.

Demand is not evenly spread (and never was)

It's tempting to think of search demand as wide and flat:

- thousands of keywords
- lots of small opportunities
- steady gains if you "cover everything"

That's not how it works.

In reality, demand clusters.

Hard.

Across local, high-intent categories like dental practices:

- A **small number of searches account for a disproportionate share of meaningful traffic**
- Everything else exists in a long tail that feels busy but converts weakly

This is not a quirk of SEO tools.

It's a reflection of human behavior.

Most people don't search creatively.

They search *efficiently*.

The illusion of progress: impressions without impact

This is where many businesses get misled.

Marketing reports often highlight:

- rising impressions
- increasing keyword counts
- broader visibility

All of that can be technically true—and strategically irrelevant.

If your visibility is growing primarily in:

- low-intent searches
- vague educational queries
- early-stage curiosity

You may feel busy without getting busier.

The system is not withholding traffic.

It's responding to **what you've aligned yourself with**.

Why "ranking for more keywords" feels logical—but fails

On paper, this sounds reasonable:

> "If we rank for more things, we'll get more traffic."

But platforms don't reward coverage.

They reward **resolution**.

Most high-value searches already have:

- established patterns
- dominant surfaces (Maps, top organic)
- clear expectations about what should appear

Ranking for dozens of marginal queries rarely adds up to the same value as being visible for a handful of decisive ones.

This is why some businesses:

- rank for hundreds of keywords
- but rely on a very small subset for actual patient acquisition

The rest are noise.

Intent concentrates traffic—and filters it

Here's the connection to the previous chapter:

High-intent searches don't just convert better.

They **attract more competition, more trust signals, and more system scrutiny**.

Which means:

- fewer businesses are allowed to appear prominently
- small ranking differences have outsized impact
- visibility jumps in steps, not slopes

This is also why:

- traffic can feel “stuck” for long periods
- then suddenly move when a threshold is crossed

The system doesn’t gradually reward effort.

It rewards **alignment that clears a confidence bar**.

Why Maps and organic absorb so much attention (conceptually)

Without getting into hard numbers yet, one pattern consistently shows up in local, trust-sensitive searches:

- Users often **scan past ads**
- They orient themselves using **Maps**
- They validate decisions using **organic listings**

This isn’t ideology.

It’s conditioning.

Over time, people learn:

- which ads feel generic
- which listings feel established
- which signals reduce regret

So while ads are visible, **they don’t automatically capture trust**.

Maps and organic often do—because they feel earned rather than rented.

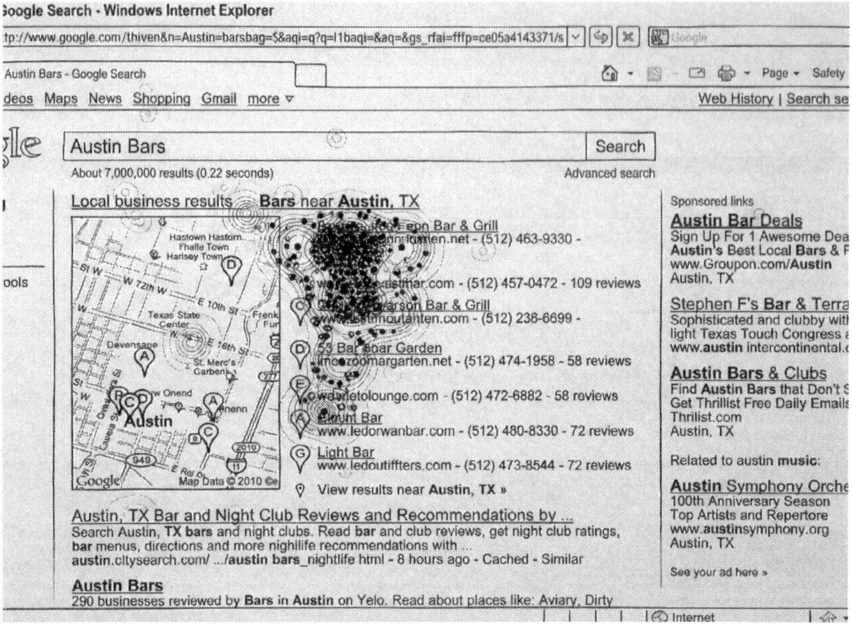

We'll quantify this later, but the key idea here is simple:

> Attention follows perceived legitimacy, not placement alone.

Volume is a poor proxy for value

Another trap is chasing search volume.

High-volume queries often:

- attract mixed intent
- sit earlier in the decision process
- produce curiosity rather than commitment

Lower-volume queries often:

- signal readiness
- cluster around action
- convert disproportionately well

But because tools emphasize volume, businesses often reverse priorities—optimizing hardest for the least decisive searches.

Again, nothing is broken.

The system is responding logically.

The interpretation is what's off.

The hidden cost of spreading effort too thin

Every page, campaign, or keyword you pursue has an opportunity cost.

When effort is spread across:

- too many topics
- too many weak signals
- too many "maybes"

You dilute the signals that matter most.

This is why some sites look active but stagnant:

- lots of content
- lots of updates
- little movement where it counts

Focus isn't a branding concept here.

It's a **signal-strength concept**.

Decision-level insight

Traffic doesn't come from "doing more."

It comes from **doing the right things in the narrow zones where demand actually concentrates**.

If your marketing feels busy but ineffective, the issue is rarely effort.

It's almost always **misalignment with where intent, trust, and competition overlap**.

In the next chapter, we'll zoom in on *why* platforms are so selective in those zones—by unpacking Google's real incentives, and why trust, not optimization, is the core currency of visibility.

CHAPTER 7

Google's Job Is Not to Help You: Trust as the Core Currency

Why this chapter matters

By this point, a quiet realization usually starts to set in.

If:

- intent drives behavior, and
- traffic concentrates around a small set of decisive searches,

then the obvious next question is:

> "Why doesn't Google just show *me* more often if I'm doing the right things?"

This chapter answers that question—and it does so by removing one of the most damaging assumptions in marketing:

That Google exists to help businesses succeed.

It doesn't.

And once you understand what it *does* exist to do, much of what feels unfair, inconsistent, or opaque starts to make sense.

Google is not neutral—and that's intentional

Google is often described as a neutral intermediary.

That's not quite right.

Google is an **opinionated system** designed to make judgment calls at scale—millions of times per second—about which outcomes are least likely to disappoint users.

Its success depends on one thing above all else:

> When people search, they trust the results enough to come back next time.

Everything else—ads, rankings, features, AI summaries—flows downstream from that requirement.

So when Google evaluates your business, it isn't asking:

- "Is this a good dental practice?"
- "Did they try hard?"
- "Is their website optimized?"

It's asking:

> "If we put this result in front of someone, how confident are we they won't regret it?"

That is a very different question.

Trust is not a feeling—it's a risk calculation

When people hear "trust," they often think about:

- branding
- tone
- bedside manner
- how something feels

Google's version of trust is colder than that.

It's probabilistic.

The system is constantly estimating:

- likelihood of relevance
- likelihood of satisfaction
- likelihood of harm or regret

Especially in categories that touch health, appearance, money, or safety.

Dental practices sit squarely in that zone.

Not as regulated as hospitals—but far more sensitive than restaurants or retail.

That positioning makes the system more cautious by default.

Why Google prefers signals over stories

You can tell a convincing story on your website.

Google can't trust stories.

At scale, stories are cheap.

Signals are not.

So the system leans heavily on:

- consistency across the web
- corroboration from third parties
- patterns of user behavior over time

Not because it's cynical—but because those signals are harder to fake *at scale* and *over time*.

This is why:

- great copy doesn't guarantee rankings
- new sites struggle even when well-built
- older, established brands often feel "sticky"

The system is not judging *intent*.

It's judging **reliability**.

Relevance, authority, experience—without the jargon

These terms get thrown around constantly, so let's strip them down to their functional meaning.

- **Relevance**
  "Is this clearly about what the person is looking for?"
- **Authority**
  "Do other credible sources treat this as legitimate?"
- **Experience**
  "Does this appear to actually serve real people in the real world?"

None of these are philosophical.

They are inference shortcuts.

Google is not verifying truth.

It's estimating confidence.

Why trust is always relative

One of the most important—and most misunderstood—ideas in SEO is that trust is **comparative**, not absolute.

You don't need to be perfect.

You need to be **more convincing than the alternatives**.

This is why:

- two dental practices can "do everything right" and still see different outcomes
- best practices alone don't guarantee movement
- optimization plateaus feel sudden and frustrating

The system isn't asking:

> "Is this good enough?"

It's asking:

> "Is this the safest choice *among these options*?"

That subtle difference explains a lot of perceived randomness.

Why effort doesn't equal reward

This is where frustration often peaks.

Owners invest time.

They invest money.

They follow advice.

And yet, visibility doesn't move.

From the system's perspective, that's not unfair—it's expected.

Effort is invisible.

Only outcomes and signals are legible.

Google can't see:

- meetings
- revisions
- strategy sessions
- intention

It can only see:

- what exists publicly
- how it aligns
- how others respond to it

This isn't unique to Google.

It's how all large-scale trust systems work.

Why this matters more for dental practices than most businesses

Because dental practices operate in a **trust-sensitive middle ground**.

- The decisions are personal
- The consequences feel permanent
- The outcomes are visible

That pushes Google to:

- favor caution

- prefer established signals
- move slowly when uncertainty exists

Which is why:

- shortcuts fail faster
- aggressive tactics backfire harder
- patience and consistency matter more than hype

This isn't a moral stance.

It's risk management.

Decision-level insight

Google is not withholding success.

It is rationing **risk**.

Once you understand that:

- SEO stops feeling arbitrary
- competition feels more logical
- trust becomes something you *build*, not demand

You don't win by convincing Google you deserve traffic.

You win by **reducing the system's perceived risk relative to your competitors**.

The next chapter explains *why* Google ended up relying on these signals in the first place—and why understanding that history helps you predict where the system goes next.

CHAPTER 8

A Short, Useful History of Search (So the Rules Make Sense)

Why this chapter matters

By now, Google's behavior probably feels *logical*—but maybe still a little rigid.

This chapter explains **why**.

Search didn't evolve to frustrate businesses.

It evolved to solve very specific problems at very large scale.

When you understand what those problems were—and how Google tried to solve them—you stop asking:

> "Why won't Google just trust us?"

And start asking:

> "What kind of system would *have* to behave this way?"

That shift is subtle, but powerful. It turns frustration into predictability.

Early search was about matching words, not judgment

In the early days of the internet, search engines were crude.

They mostly did one thing:

- Look for pages that repeated the same words the user typed

That worked—until it didn't.

People quickly learned they could:

- stuff pages with keywords
- hide text
- repeat phrases unnaturally

And suddenly, search results became noisy, misleading, and low quality.

The problem wasn't technical.

It was **incentive-driven**.

PageRank changed the question

Google's breakthrough wasn't better text matching.

It was a different question entirely:

> "What if we judged pages by how the rest of the web treated them?"

Instead of asking:

- "Does this page say the right words?"

Google asked:

- "Do other pages *vouch* for this one?"

Links became a proxy for trust—not because they were perfect, but because they were:

- harder to fake at scale
- external to the site owner
- accumulated over time

This mattered enormously.

For the first time, search wasn't just indexing content.

It was **inferring credibility**.

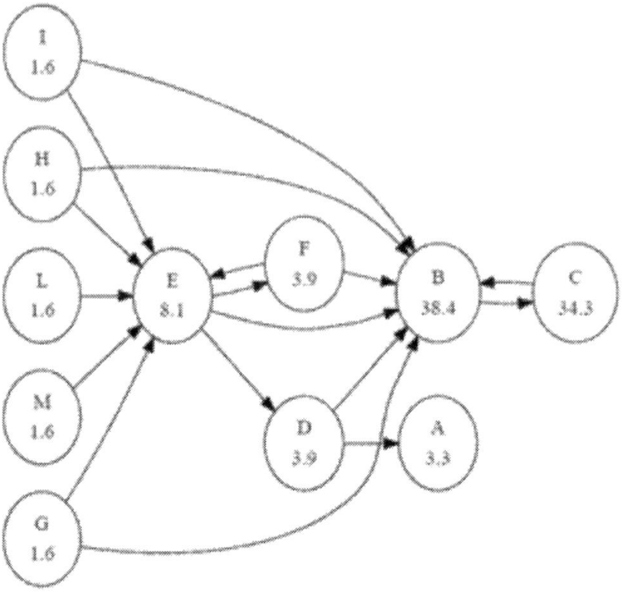

Why links worked—and why they got abused

Links worked because they represented *friction*.

Someone had to:

- discover you
- decide you were worth referencing
- associate their reputation with yours

But once people understood that links mattered, they started chasing them—not earning them.

This led to:

- link farms
- paid links

- reciprocal schemes

Again, not because marketers were unethical—but because incentives rewarded manipulation.

The system had to respond.

Why algorithms kept getting more complex

Each wave of abuse forced Google to ask:

> "What signal is harder to fake?"

That's why over time, the system began layering in:

- context around links
- patterns of behavior
- consistency across sources
- signals from Maps, reviews, and user interaction

Not to punish businesses—but to **reduce false confidence**.

This is also why:

- single tactics stop working
- shortcuts decay quickly
- systems reward breadth of credibility over cleverness

Complexity wasn't the goal.

Resilience was.

Why “white hat vs. black hat” is the wrong framing

This distinction gets oversimplified.

It’s not really about:

- good vs bad
- ethical vs unethical

It’s about **alignment vs exploitation**.

Tactics that align with the system’s goals:

- scale more slowly
- compound more reliably
- survive updates

Tactics that exploit gaps:

- work briefly
- fail suddenly
- create instability

That pattern has repeated for over two decades—and it’s why long-term visibility always favors consistency over cleverness.

Why this history matters *now*, especially with AI

Here’s the critical connection.

AI did not change Google’s core problem.

It **intensified it**.

If content becomes easier to produce:

- volume increases
- noise increases
- manipulation becomes cheaper

Which means the system must lean even harder on:

- trust signals
- corroboration
- real-world behavior

This is why AI doesn't replace trust—it makes it more valuable.

The same forces that shaped search from the beginning are still at work.

They're just operating faster.

Why dental practices feel this pressure more acutely

Because your category:

- affects people personally
- carries visible consequences
- involves perceived risk

When risk goes up, systems tighten.

They always have.

So when owners feel like:

> “The bar keeps getting higher”

What’s actually happening is:

> “The system is protecting itself from being wrong.”

That’s not personal.

It’s structural.

Decision-level insight

Search didn’t become complex to exclude you.

It became complex to **protect trust at scale**.

Every major change—from links to Maps to AI summaries—follows the same logic:

- reduce regret
- increase confidence
- reward corroborated reality

When you understand that history, you stop chasing tactics that fight the system—and start investing in signals that *survive* it.

The next chapter brings this history into the present by unpacking the most misunderstood—and most powerful—local surface for dental practices: **Maps**.

CHAPTER 9

Maps: Where Trust Becomes Local

Why this chapter matters

For most dental practices, Maps is where marketing becomes real.

It's no longer abstract visibility or long-term brand building. It's the moment a patient is standing still, phone in hand, deciding where to go. Not tomorrow. Not eventually. Now.

That immediacy is why Maps is often misunderstood.

Owners treat it like a ranking contest when, in reality, Maps is a confidence system—designed to minimize regret, not reward effort.

This chapter exists to explain how that confidence is formed, why being "number one" is often irrelevant, and why reviews have quietly become the strongest local signal you control.

Maps does not rank businesses — it ranks confidence

The most important thing to understand about Google Maps is this:

Google is not ranking dental practices.

Google is ranking its confidence in recommending them.

That confidence answers three questions:

- Does this business exist and operate consistently?
- Does it deliver what it claims to offer?
- Will a patient likely have a satisfactory experience?

Everything Maps shows—the listings, the order, the radius—flows from those questions.

This is why Maps behaves differently than organic search or ads. It is not primarily trying to discover the best option. It is trying to avoid the worst outcome.

Why proximity matters less than most people think

Many owners still believe proximity is the dominant Maps factor.

That was true years ago. It is no longer the full picture.

Modern Maps behavior increasingly favors prominence and trust signals over raw distance. A dental practice that is slightly farther away—but consistently reviewed, well-aligned, and clearly legitimate—can and often does outrank closer competitors with weaker signals.

This shift reflects how Google itself has changed:

- AI systems summarize authority, not distance
- Voice search requires a single confident answer
- Spam suppression penalizes address manipulation

Maps is adapting accordingly.

Proximity still matters—but it is no longer the deciding factor owners assume it is.

Why being in the top three is just as good as being number one

Another persistent misunderstanding is the obsession with being "#1 on Maps."

From a patient's perspective, Maps does not feel like a leaderboard.

When someone searches locally, their behavior typically looks like this:

1. They see the top three listings.
2. They scan reviews, ratings, and recency.
3. They click into one or two profiles.
4. They choose based on perceived trust.

Very few patients choose based on micro-rank position.

Being in the top three means you've already passed Google's confidence threshold. From there, reviews decide selection, not position.

A well-reviewed third listing routinely outperforms a poorly reviewed first listing—both in clicks and calls.

Maps surfaces options.

Reviews determine the choice.

Maps visibility is geographic, not absolute

One reason Maps feels inconsistent is because it is location-dependent.

Your visibility changes block by block, not city by city. A dental practice can appear highly visible near its core area and fade gradually farther out—without anything being "wrong."

This isn't instability. It's contextual confidence.

One way some owners orient themselves is by visualizing visibility across a service area using pin-based Maps ranking grids. These don't explain why rankings exist—but they do show where confidence holds and where it drops off.

If you want to see how your Maps visibility actually behaves geographically, you can access a free local Maps ranking report at https://patientflow.us/resources-dental.

The value isn't optimization.

It's clarity.

Reviews are not marketing — they are trust infrastructure

Reviews are often discussed as a conversion tactic.

That framing is too small.

From Google's perspective, reviews are evidence:

- that real patients are being served,
- that outcomes are generally satisfactory,
- that the business is active and operational,
- and that expectations align with reality.

This is why review velocity and consistency matter more than raw totals.

A dental practice with steady, ongoing feedback signals health. A business with long gaps in reviews signals uncertainty—regardless of how good past reviews were.

Maps interpret silence as risk.

Why review volume beats perfect ratings

Many owners worry about maintaining a flawless rating.

Google doesn't expect perfection. Patients don't either.

What matters more is:

- a meaningful number of reviews,
- steady recency,
- realistic distribution,
- and active responses.

A business with hundreds of reviews averaging 4.6 looks more trustworthy than one with 12 reviews averaging 5.0.

Volume signals experience.

Experience signals safety.

Getting reviews is an operational system, not a campaign

The most successful dental practices don't "run review campaigns."

They build review requests into patient flow.

This matters, because Google is very good at detecting artificial spikes. Sudden bursts followed by silence look unnatural. Consistency looks real.

Healthy review profiles are created when:

- asking is normalized,
- timing is intentional,
- and the request feels like feedback, not a favor.

A practical framework for earning more reviews (without pressure)

This is one of the few places where guidance is warranted, because reviews sit at the intersection of marketing and operations.

1. Make reviews routine

Reviews should feel expected, not exceptional.

Patients are more willing when it feels like standard feedback.

2. Ask at the right moment

The best time to ask is when satisfaction is visible—often immediately after treatment or a successful outcome.

3. Reduce friction completely

Direct links, QR codes, or simple follow-ups matter. Friction kills participation.

4. Train staff, not scripts

Staff should understand why reviews matter and ask naturally. Awkwardness is usually a systems issue, not a personality issue.

5. Respond to every review

Responses signal engagement, accountability, and ongoing management—to both patients and Google.

Reviews reinforce the entire marketing system

Reviews don't live only on Maps.

They:

- validate website claims,
- reduce ad friction,
- influence AI summaries,
- shape patient expectations before contact,
- and reinforce local authority across platforms.

They are one of the few signals that compound everywhere.

That makes them infrastructure, not tactics.

Decision-level insight

Maps does not reward effort.

It rewards confidence at scale.

Being in the top three is enough—if reviews support selection.

Visibility expands geographically as trust accumulates.

Reviews compound when they are embedded operationally.

If Maps feels frustrating, the issue is rarely rank alone.

It's whether Google has enough evidence to commit.

The next chapter moves from Maps to the website—the place where that trust is either reinforced or quietly undermined.

CHAPTER 10

The Website: Structural Clarity and Credibility

Why this chapter matters

For most dental practice owners, the website feels like the obvious centerpiece of marketing.

It's where ads send traffic.

It's where social profiles point.

It's what SEO companies talk about the most.

And yet, it's also where a lot of marketing quietly fails—not because the website is "bad," but because it's trying to do too many jobs at once.

This chapter exists to reset how you think about your website.

Not as a brochure.

Not as a design project.

Not as an SEO checklist.

But as the **structural backbone** that makes every other signal—Maps, content, social, ads—legible and believable to the system.

The website is not a channel—it's the source of truth

Here's the simplest reframing that unlocks everything else:

> Your website is not one marketing channel among many.
>
> It is the reference point that all other channels are compared against.

Maps listings, ads, social profiles, citations, and even AI systems are constantly trying to answer one quiet question:

> "Does this business present a clear, consistent reality everywhere?"

The website is where that reality is supposed to be spelled out.

When it isn't, everything downstream becomes weaker—even if those other channels are "optimized."

Structure matters more than content volume

Most website advice focuses on *what* to add:

- more pages
- more blog posts
- more words

But platforms care far more about **how information is organized** than how much of it exists.

Structure is how systems understand:

- what you actually offer
- which services matter most
- how locations relate to each other
- what the business specializes in versus what it merely mentions

If structure is unclear, content doesn't clarify—it confuses.

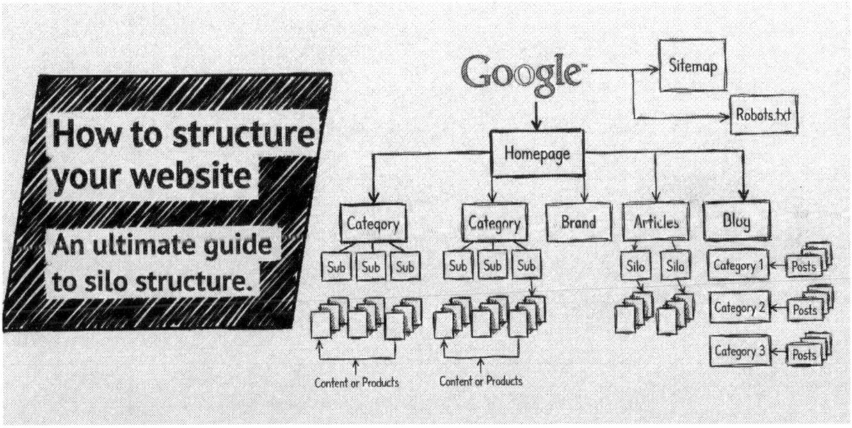

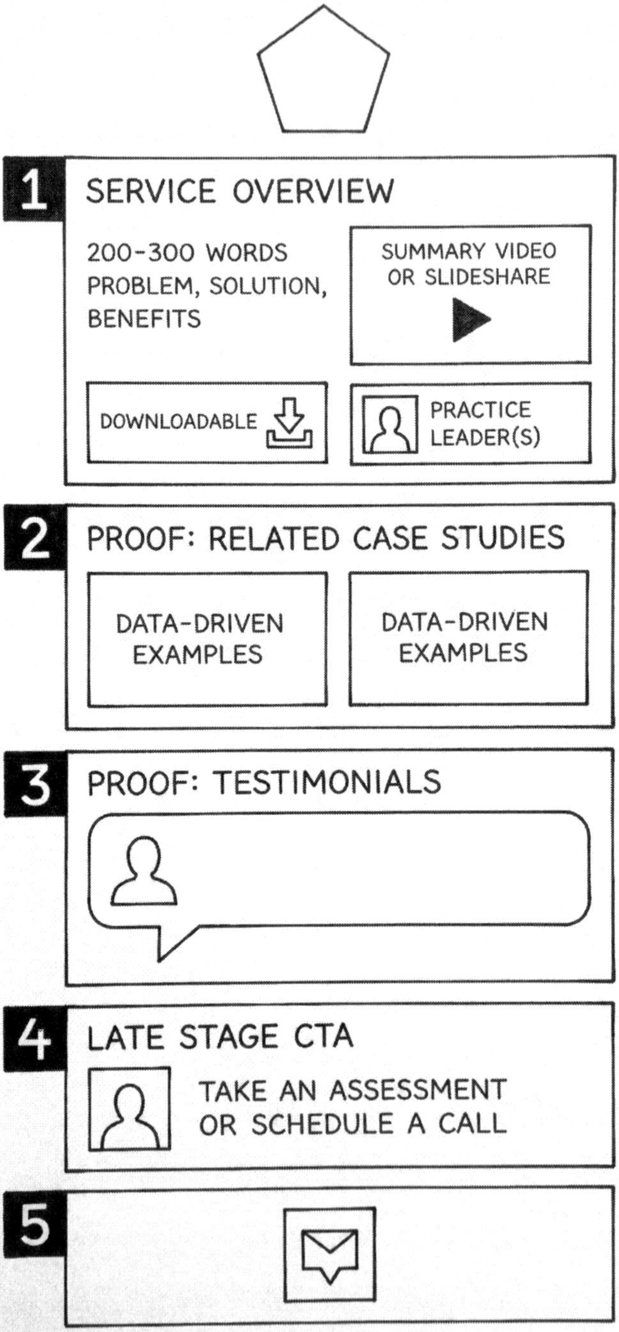
1
SERVICE OVERVIEW
200-300 WORDS
PROBLEM, SOLUTION,
BENEFITS
SUMMARY VIDEO
OR SLIDESHARE
DOWNLOADABLE
PRACTICE
LEADER(S)
2
PROOF: RELATED CASE STUDIES
DATA-DRIVEN
EXAMPLES
DATA-DRIVEN
EXAMPLES
3
PROOF: TESTIMONIALS
4
LATE STAGE CTA
TAKE AN ASSESSMENT
OR SCHEDULE A CALL
5

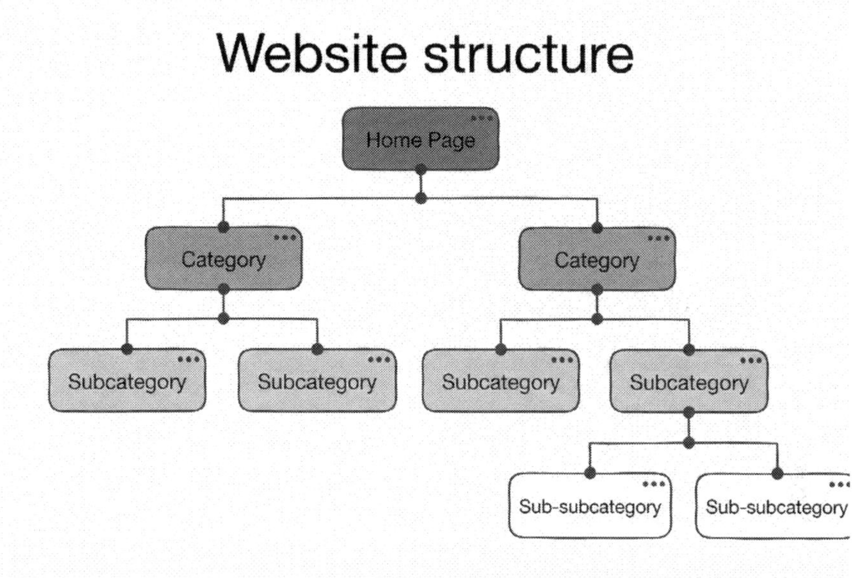

Why clarity beats cleverness every time

Dental practices are especially vulnerable to this problem because:

- services overlap conceptually
- language varies by provider
- trends introduce new terminology constantly

It's tempting to sound sophisticated.

It's tempting to consolidate everything into broad, elegant pages.

But systems don't reward elegance.

They reward **unambiguous meaning**.

A site that is:

- plain
- repetitive
- structurally obvious

often outperforms a site that is:

- beautifully written
- conceptually dense
- creatively organized

This isn't because Google prefers boring websites.

It's because boring is easier to **understand at scale**.

Service pages are not sales pages

One of the most common misunderstandings is treating service pages like conversion tools.

They aren't.

Their primary job is not persuasion.

It's **definition**.

A service page exists to answer:

- What is this service?
- Who is it for?
- Where is it offered?
- How does it relate to other services?

Conversion happens later—often after:

- Maps validation
- review scanning
- social reinforcement
- return visits

When service pages try to do everything at once, they usually fail at the one job the system actually needs them to do.

Location pages are about legitimacy, not reach

Location pages are another area where intent gets misread.

They're often created to "rank in more cities."

But their real function is to:

- establish physical relevance
- support Maps signals
- clarify service availability

When done well, they help the system confidently say:

> "This business really operates here."

When done poorly, they dilute trust and introduce doubt.

Again, structure matters more than ambition.

Why siloing isn't an SEO trick—it's a comprehension tool

"Siloing" gets talked about like a tactic.

It's not.

It's simply a way of grouping related information so that:

- humans can follow it
- systems can summarize it
- AI can synthesize it

When services, locations, content, and navigation reinforce each other, the site stops feeling like a collection of pages and starts feeling like a **coherent entity**.

That coherence is a trust signal.

The website as the hub of alignment

Here's where the holistic model starts to click.

Your website should:

- reflect what Maps claims
- support what ads promise
- reinforce what social posts show
- anchor what blog content explores

When those things disagree—even subtly—the system doesn't argue with you.

It just hesitates.

And hesitation is the enemy of visibility.

Why redesigns rarely fix structural problems

Many owners sense something is "off" with their website and assume the answer is:

- a redesign
- new copy
- a fresh look

But if the underlying structure doesn't change, the results usually don't either.

You've just repainted the same map.

The issue was never aesthetics.

It was **clarity**.

Decision-level insight

A strong dental practice website doesn't try to impress.

It tries to be **understood**.

It exists so that when Google, Maps, social platforms, ad systems, and AI models all look for confirmation, they find the same story —told plainly, consistently, and without contradiction.

When your website plays that role well, every other channel gets stronger.

Many owners suspect their website is part of the problem but struggle to name why. If you want a clearer picture, you can get a website clarity rating at **https://patientflow.us/resources-dental** to evaluate structure, messaging alignment, and credibility signals—not design preferences.

The next chapter builds directly on this by explaining how **content and blogging** support that structure over time—not as traffic bait, but as ongoing proof of expertise and alignment.

CHAPTER 11

Content and Blogging: Demonstrated Expertise Over Time

Why this chapter matters

Content is where good intentions go to die.

Dentists are constantly told:

- "You need blogs for SEO."
- "Content builds authority."
- "Posting consistently is important."

So they publish.

They stay active.

They fill calendars.

And yet, very often, nothing meaningful changes.

This chapter exists to explain **why content so often feels busy but ineffective**—and how to reframe blogging and educational content as something far more valuable than traffic bait.

Done correctly, content isn't there to attract everyone.

It's there to **prove consistency, reinforce structure, and support trust over time**.

Content is not a growth lever by itself

This is the uncomfortable truth:

> Most blog content does not drive patient acquisition directly.

That doesn't mean content is useless.

It means its role is misunderstood.

Content rarely convinces someone to choose you *on its own*.

What it does instead is quietly answer follow-up questions after trust has already begun forming.

In other words:

- Content doesn't start most journeys
- It **stabilizes** them

That distinction changes everything about how it should be created and evaluated.

Blogging is how expertise becomes legible

Expertise isn't declared.

It's inferred.

The system doesn't ask:

> "Is this business knowledgeable?"

It asks:

> "Do they consistently explain things the way knowledgeable businesses do?"

That's where blogging matters.

Not because one article ranks.

But because **patterns accumulate**:

- consistent topics
- consistent explanations
- consistent alignment with services offered

Over time, content answers a quiet question for the system:

> "Does this business regularly help people understand decisions in this category?"

That's a trust signal.

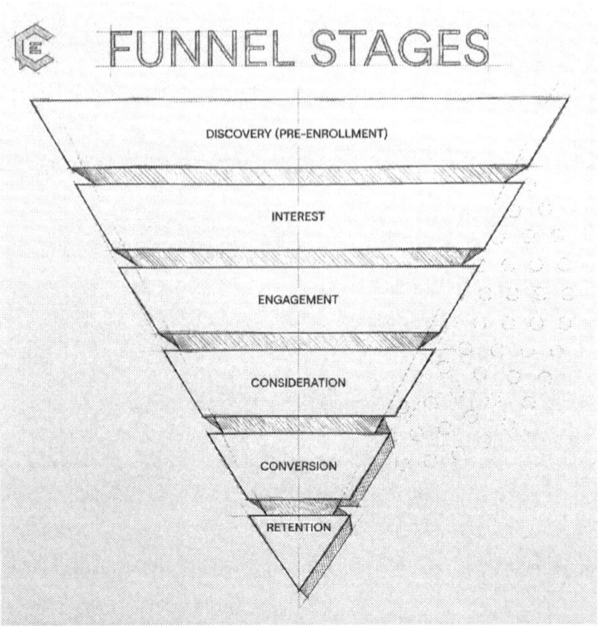

Anatomy of a Content Cluster

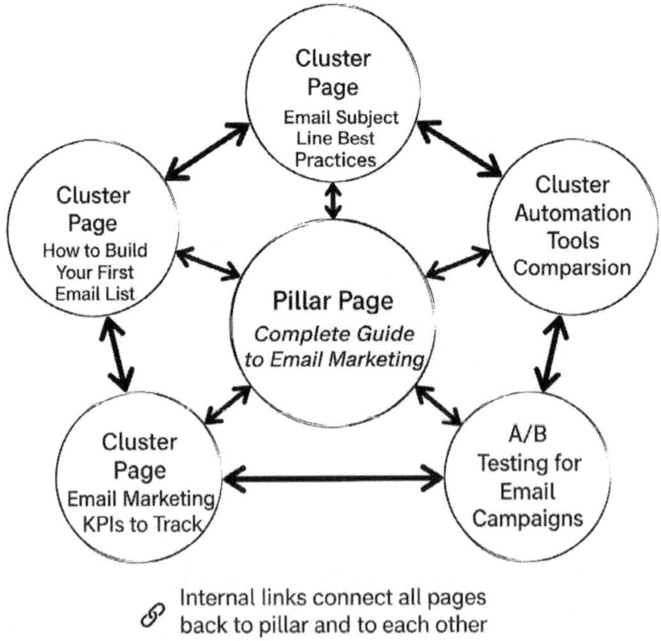

Why random blog topics weaken trust

One of the biggest mistakes businesses make is treating blogs as creative outlets.

Posts drift into:

- trends
- tangents
- loosely related wellness topics
- seasonal filler

Individually, none of this is harmful.

Collectively, it creates **topical ambiguity**.

Remember: the system is trying to summarize who you are.

When content wanders, the summary gets fuzzy.

Blogs work best when they:

- reinforce core services
- answer adjacent questions
- deepen understanding of what you already offer

Content should make your site *more specific*, not more interesting.

Content supports structure—it doesn't replace it

This is where Chapter 10 and Chapter 11 lock together.

Your service pages define **what you do**.

Your content explains **how and why those services matter**.

When blogs:

- link naturally back to core service pages
- expand on real patient questions
- use consistent language

They strengthen the site's overall clarity.

When they don't, they act like static noise.

Content is not an alternative to structure.

It's reinforcement.

Why "educational" doesn't mean "early-stage only"

There's a common assumption that educational content is only for people far from making a decision.

In reality, even high-intent patients often:

- read explanations to validate choices
- skim risks and expectations
- look for alignment with their understanding

Educational content doesn't just attract beginners.

It **reduces hesitation** for people already close to acting.

That's why tone matters.

This isn't about teaching from scratch.

It's about confirming competence.

AI changes content volume—but raises the bar for coherence

AI makes content easier to produce.

That does *not* mean content matters less.

It means **disorganized content gets exposed faster**.

When AI systems analyze your site, they don't read one article.

They infer themes across *all* of them.

Which means:

- consistent content compounds
- scattered content contradicts itself
- thin content becomes obvious

AI doesn't reward novelty.

It rewards **pattern clarity**.

Blogs are long-term signals, not short-term campaigns

Content works slowly because trust works slowly.

That's not a flaw.

That's protection.

A blog strategy that:

- reinforces the same services
- answers the same categories of questions
- stays aligned month after month

does something powerful:

It tells the system you are **still the same business**, still serving the same needs, still relevant.

Stability is a signal.

Decision-level insight

Blogging isn't about publishing more.

It's about **saying the same important things clearly, over time, in slightly different ways**.

When content:

- aligns with your website structure,
- reinforces your Maps presence,
- supports what ads and social already claim,

it becomes cumulative proof—not marketing noise.

Before producing more content—especially with AI—it helps to define what your content should reinforce at all. Some readers use structured content strategy frameworks, like those available at https://patientflow.us/resources-dental, to clarify themes and boundaries before writing anything new.

The next chapter explains the opposite side of this equation:

off-site signals—the trust you *don't* control, but which ultimately decide how credible everything else looks.

CHAPTER 12

Off-Site Signals: Authority You Don't Control

Why this chapter matters

Up to this point, everything we've discussed lives largely inside your control.

- Your website structure
- Your service pages
- Your content
- Your messaging consistency

This chapter introduces the uncomfortable counterweight to all of that:

> **The system trusts what others say about you more than what you say about yourself.**

For dental practices, this is where reality either confirms your story—or quietly undermines it.

Why off-site signals exist at all

From Google's perspective, relying only on a business's own website would be reckless.

Anyone can:

- write polished copy
- claim expertise
- describe themselves favorably

So the system asks a harder question:

> "Do *other* sources describe this business in the same way?"

Off-site signals exist to answer that question.

They are not about marketing flair.

They are about **corroboration**.

What counts as an off-site signal (at a high level)

Off-site signals include things like:

- business citations and listings
- reviews and ratings
- brand mentions
- links from other websites
- consistency across directories

Individually, none of these are decisive.

Collectively, they form a pattern.

Google isn't counting points.

It's assessing **agreement**.

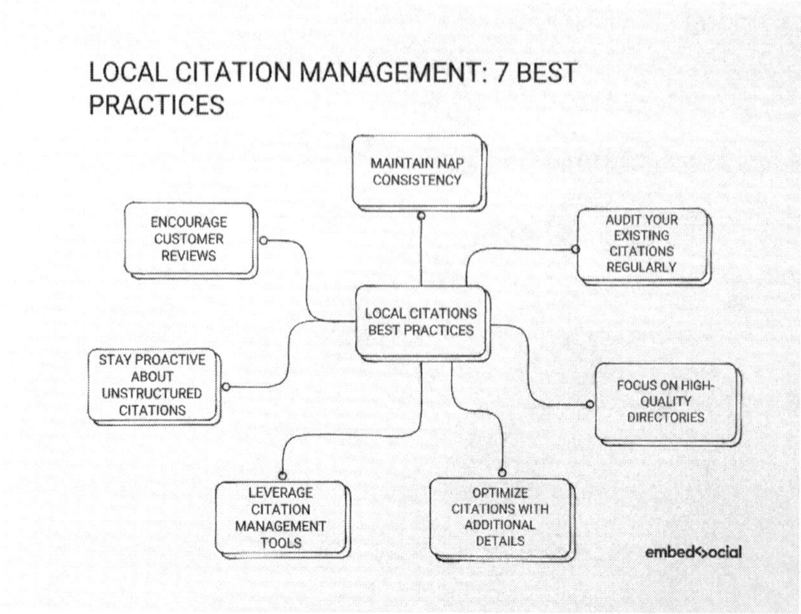
LOCAL CITATION MANAGEMENT: 7 BEST PRACTICES
MAINTAIN NAP CONSISTENCY
ENCOURAGE CUSTOMER REVIEWS
AUDIT YOUR EXISTING CITATIONS REGULARLY
LOCAL CITATIONS BEST PRACTICES
STAY PROACTIVE ABOUT UNSTRUCTURED CITATIONS
FOCUS ON HIGH-QUALITY DIRECTORIES
LEVERAGE CITATION MANAGEMENT TOOLS
OPTIMIZE CITATIONS WITH ADDITIONAL DETAILS
embed<>ocial

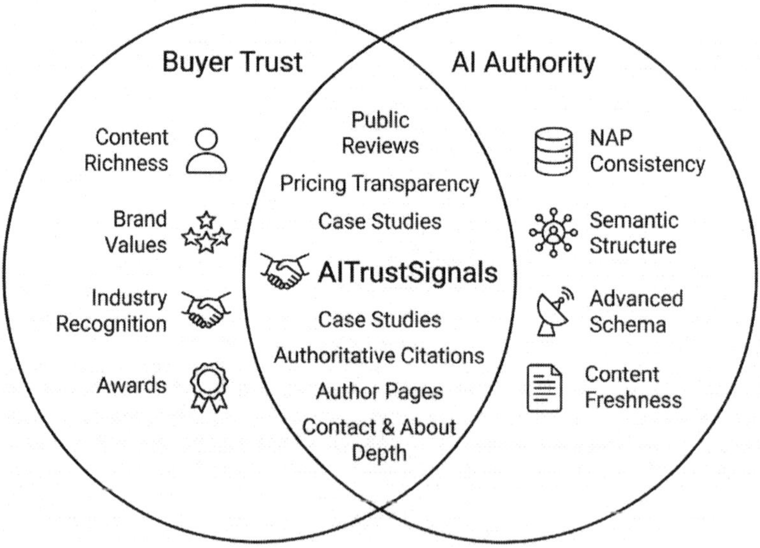
Buyer Trust
AI Authority
Content Richness
Brand Values
Industry Recognition
Awards
Public Reviews
Pricing Transparency
Case Studies
AITrustSignals
Case Studies
Authoritative Citations
Author Pages
Contact & About Depth
NAP Consistency
Semantic Structure
Advanced Schema
Content Freshness

Why consistency matters more than volume

One of the most common misunderstandings is assuming off-site SEO is about *quantity*.

More links.

More listings.

More reviews.

But the system is less impressed by volume than by **alignment**.

When:

- your business name varies
- your categories drift
- your services don't match
- your locations blur

The system hesitates.

Not because anything is "wrong," but because the story becomes harder to trust.

Consistency reduces cognitive load.

For machines and humans.

Citations are about existence, not prestige

Citations—mentions of your business name, address, and phone—are often dismissed as boring.

That's exactly why they matter.

They answer a very basic question:

> "Does this business exist in the real world the way it claims to?"

For local, trust-sensitive categories like dental practices, existence is not assumed.

It's inferred.

Citations don't make you impressive.

They make you **plausible**.

And plausibility is the floor on which everything else rests.

Reviews are trust signals, not reputation management

Reviews get emotionally charged because they feel personal.

From the system's perspective, they're functional.

Reviews help answer:

- Are people actually choosing this business?
- Do experiences align with expectations?
- Is engagement ongoing or sporadic?

The content of reviews matters—but so does the pattern:

- steady accumulation vs sudden spikes
- specificity vs generic praise

- alignment with stated services

Reviews aren't there to flatter you.

They're there to **reduce regret**.

Why links still matter—even if they're misunderstood

Links are often treated as an outdated SEO tactic.

They aren't.

They're just misunderstood.

Links work because they represent:

- editorial choice
- contextual association
- reputational risk

Someone else decided to associate their presence with yours.

That's expensive at scale.

Which is why links remain useful signals—even as the system becomes more sophisticated.

What's changed is *how* they're interpreted:

- context matters more than raw count
- relevance outweighs novelty
- patterns matter more than individual wins

Off-site signals don't override weak foundations

This is critical.

Off-site signals **reinforce** clarity.

They do not compensate for its absence.

If:

- your website is vague
- your services are unclear
- your messaging drifts

Strong off-site signals don't fix that.

They amplify confusion.

The system can't reconcile disagreement between what you say and what others imply.

So it chooses caution.

Why off-site work feels slow (and why that's good)

Off-site trust accumulates gradually.

That frustrates owners—but it protects users.

Trust that can be built overnight can be manipulated overnight.

Trust that requires time resists shortcuts.

This is why:

- aggressive campaigns decay

- manufactured signals lose value
- stable patterns outperform bursts

The system is not punishing speed.

It's privileging **reliability**.

Decision-level insight

Off-site signals are not levers you pull.

They are **votes of confidence you earn**—often indirectly.

Your job isn't to chase them aggressively.

It's to make sure that when they appear, they **agree with everything else**:

- your Maps presence
- your website structure
- your content themes
- your real-world behavior

When off-site signals reinforce your story, trust compounds.

When they contradict it, growth stalls quietly.

The next chapter explains a channel many businesses misunderstand in this exact way—**social media**—and why its real value isn't traffic, but reinforcement.

CHAPTER 13

Social Media: Trust Echoes, Not a Traffic Engine

Why this chapter matters

Social media is where a lot of dental practice marketing becomes quietly misaligned.

Not because social media is ineffective—but because it's almost always asked to do the *wrong job*.

Owners are told:

- "Social drives awareness."
- "You need to post consistently."
- "Engagement builds your brand."

So feeds get filled.

Content gets scheduled.

Metrics get tracked.

And yet, very few dental practices can point to social media and say, *"That's where our patient acquisition really comes from."*

This chapter exists to reframe social media's role—not as a lead engine, but as something far more subtle and just as important: **a trust echo**.

Social media rarely creates demand—it reflects it

For most dental practices, social media is not where demand begins.

People don't usually decide they want straight teeth *because* of an Instagram post. That desire forms elsewhere—through life events, conversations, exposure, and timing.

What social media does instead is answer a different question:

> "Is this business active, real, and aligned with what I expect?"

In that sense, social media behaves less like advertising and more like **confirmation**.

It doesn't convince.

It reassures.

Why social platforms matter to search systems (even indirectly)

Here's the important systems-level insight:

Search engines don't need to crawl every social post to feel their effect.

They infer reality from **patterns across the web**.

When a business:

- posts consistently about its real services
- uses the same language as its website
- shows real people, real environments, real outcomes
- stays aligned month after month

That consistency becomes part of the broader trust landscape—even if no single post "ranks."

Social media contributes to the *believability* of everything else.

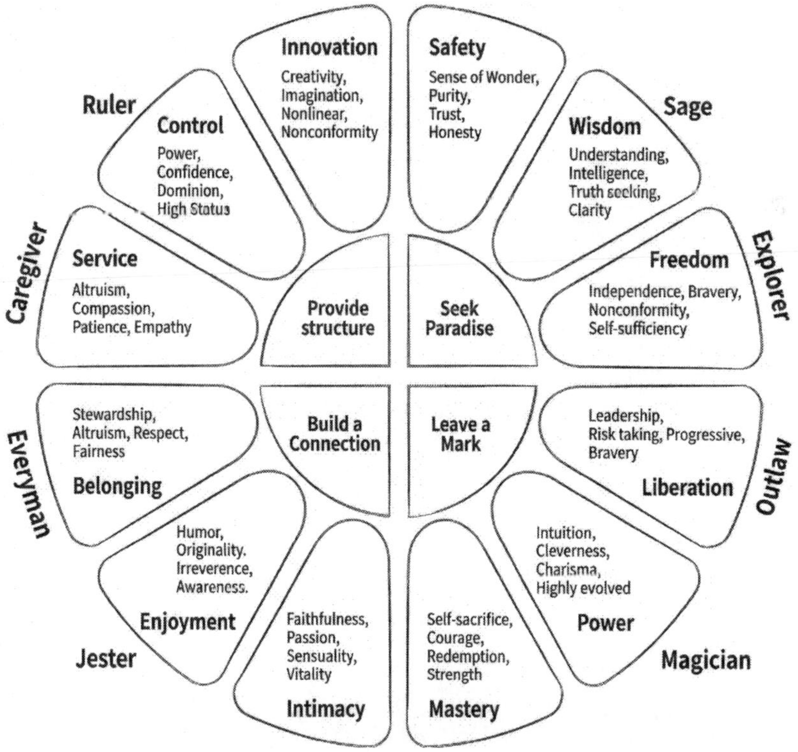

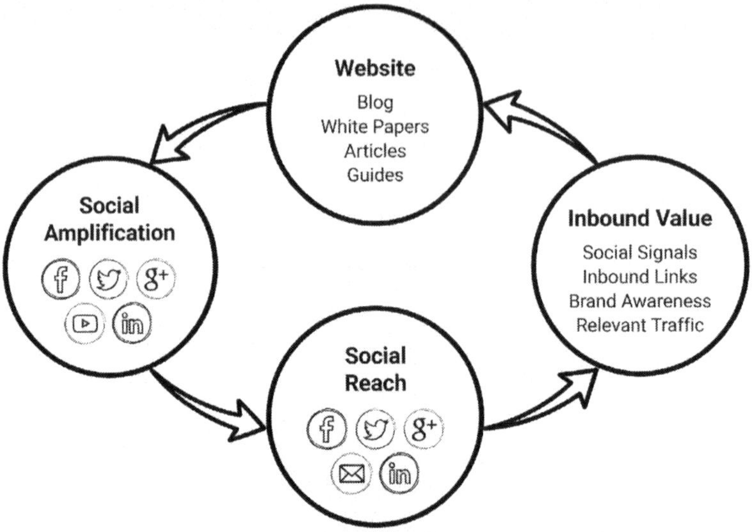

Why disconnected social content actively hurts clarity

One of the most common mistakes is treating social media as a creative sandbox.

Posts drift into:

- generic wellness quotes
- trending audio with no relevance
- lifestyle content unrelated to services
- promotions that don't match the website

Individually, these feel harmless.

Systemically, they create noise.

Remember: the system is constantly trying to answer:

> "What kind of business is this, really?"

When social media tells a different story than:

- your website
- your Maps profile
- your service pages
- your reviews

The result isn't punishment.

It's hesitation.

And hesitation slows everything.

Social works best when it repeats—not invents

This is counterintuitive for many businesses.

We're taught that social media rewards novelty and creativity.

Platforms might—but **trust systems don't**.

From a systems perspective, social media is most powerful when it:

- reinforces core services
- mirrors website language
- echoes blog themes
- visually validates Maps listings

Repetition is not redundancy.

It's confirmation.

What feels boring internally often looks **stable and trustworthy** externally.

Why engagement metrics are the wrong scorecard

Likes, shares, and comments are easy to track—and easy to overvalue.

They feel like feedback.

But for most dental practices:

- high engagement doesn't correlate cleanly with patient acquisition
- viral posts rarely align with high-intent demand
- quiet, consistent posting often outperforms flashy experiments over time

This doesn't mean engagement is meaningless.

It means it's **secondary**.

The real question isn't:

> "Did this post perform?"

It's:

> "Does this post support the same story everything else is telling?"

Social media as visual corroboration

Social does one thing better than almost any other channel:

It shows *reality*.

- the space
- the staff
- the tone
- the professionalism
- the consistency

That visual evidence matters in trust-sensitive categories.

When someone:

- sees you in Maps,
- visits your website,
- reads reviews,
- then checks social,

they're not looking for novelty.

They're looking for **alignment**.

When they find it, confidence increases.

When they don't, doubt creeps in.

AI increases the cost of inconsistency

AI systems don't "like" social media.

They **summarize patterns** across it.

As AI becomes better at synthesizing brand presence:

- inconsistent messaging becomes easier to detect
- scattered themes become more obvious
- disconnected posts stop being invisible

AI doesn't require perfection.

It requires coherence.

Social media either supports that—or undermines it faster than ever before.

Decision-level insight

Social media is not where most dental practice decisions begin.

But it's often where they're **confirmed or rejected**.

When social content:

- aligns with your website,
- reinforces your services,
- echoes your expertise,
- reflects your real-world presence,

it becomes a quiet but powerful trust multiplier.

When it doesn't, it weakens everything else—slowly, silently, and without obvious warning.

The next chapter ties all of this together by explaining **why marketing works when signals agree**, and why coherence—not effort—is the real competitive advantage.

CHAPTER 14

Why Marketing Works When Signals Agree

Why this chapter matters

Up to now, we've talked about channels.

- Maps
- Website
- Content
- Off-site signals
- Social media

Individually, each of these makes sense. Each has rules. Each has logic.

But this chapter is where the book **locks together**.

Because in the real world—and especially in a post-AI search landscape—marketing doesn't succeed because one channel is strong. It succeeds because **nothing contradicts anything else**.

This chapter explains why coherence beats optimization, why alignment compounds, and why most stalled marketing isn't underpowered—it's fragmented.

The system is not evaluating channels—it's inferring reality

Here's the most important reframing in the entire book:

> Google is not scoring your SEO, your Maps, your social media, and your ads separately.
>
> It is inferring **what kind of business you are** from all of them at once.

Each surface is a signal.

But signals don't stand alone.

They **corroborate—or conflict**.

The system is constantly asking:

- Do these sources agree?
- Do they tell the same story?
- Do they describe the same business, in the same way, over time?

When the answer is yes, confidence increases.

When the answer is no, everything slows down.

Not as punishment.

As caution.

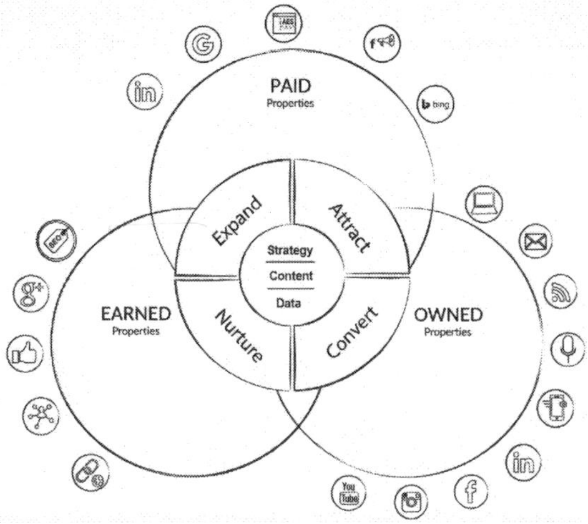
PAID
Properties
EARNED
Properties
OWNED
Properties
Expand
Attract
Nurture
Convert
Strategy
Content
Data

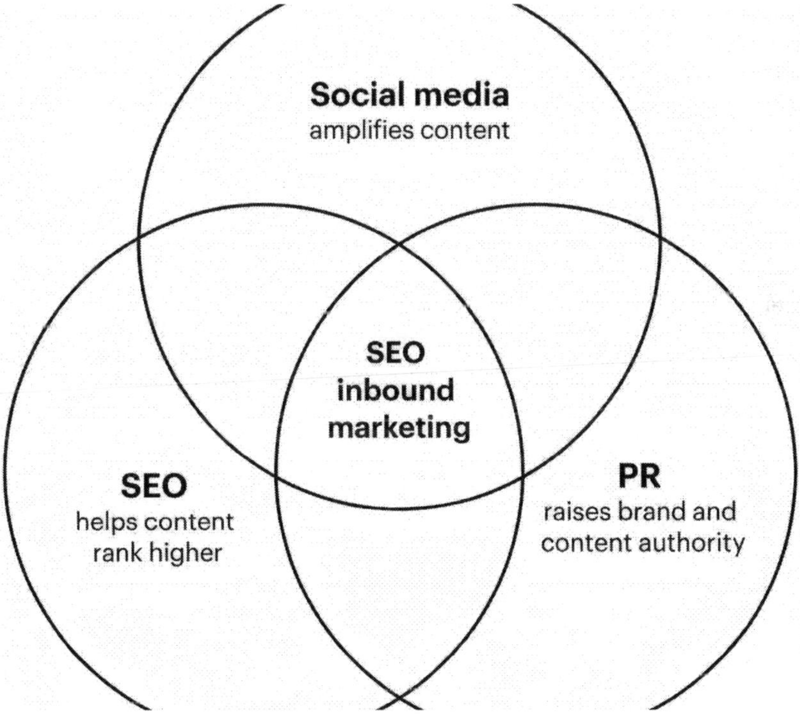
Social media
amplifies content
SEO
inbound
marketing
SEO
helps content
rank higher
PR
raises brand and
content authority

How contradiction quietly kills momentum

Most marketing breakdowns aren't dramatic.

They don't look like penalties or crashes.

They look like:

- rankings that stall
- Maps visibility that fluctuates
- ads that get more expensive
- content that "doesn't seem to help"

Often, the root cause is simple:

> The signals don't agree.

For example:

- Maps categories don't match service page emphasis
- Social content highlights trends your website barely mentions
- Blog topics drift away from core offerings
- Ads promise things your site doesn't clearly support

None of these are catastrophic on their own.

Together, they introduce doubt.

And doubt is what the system avoids distributing.

Alignment reduces risk—for everyone

Remember Google's job.

It's not to help you grow.

It's to reduce regret.

Aligned signals do exactly that.

When:

- your website structure matches your Maps profile
- your content reinforces your services
- your social posts reflect real expertise
- your off-site mentions agree with your positioning

The system can make a safer decision.

It doesn't have to *interpret* you.

It can *recognize* you.

Recognition scales.

Interpretation slows things down.

Why holistic systems outperform "strong channels"

This is where many businesses get stuck.

They say:

> "But our website is good."
>
> "But our social engagement is high."
>
> "But our ads convert."

Those statements can all be true—and still fail to produce durable growth.

Because strength in isolation doesn't compound.

Holistic systems do.

When channels reinforce each other:

- gains accumulate faster
- volatility decreases
- small improvements matter more

That's why some dental practices:

- seem to "break through" suddenly

- feel easier to grow once momentum starts
- resist algorithm changes better than others

They aren't lucky.

They're coherent.

Why AI makes alignment non-negotiable

AI doesn't evaluate marketing the way humans do.

It doesn't admire cleverness.

It doesn't reward novelty.

It doesn't overlook inconsistency.

AI systems summarize patterns.

They look for:

- repetition
- agreement
- stability
- thematic clarity

Which means:

- aligned systems scale faster
- fragmented systems unravel faster

AI doesn't create winners.

It **amplifies structure**.

That's why the gap between "well-run" and "chaotic" marketing is widening—not narrowing.

This is why "doing more" stops working

When marketing feels stalled, the instinct is to add:

- another channel
- another campaign
- another vendor
- another tool

But adding volume to a misaligned system increases noise, not clarity.

The better question is:

> "Do all of our signals currently agree about who we are and what we do?"

If the answer is no, adding more effort usually makes results worse—not better.

What alignment actually looks like (at a high level)

Without turning this into a checklist, aligned dental practice marketing tends to share a few characteristics:

- Core services are emphasized everywhere, not just on the website

- Language is consistent across Maps, content, social, and ads
- Educational content supports services instead of wandering
- Visuals match real-world experience
- Messaging stays stable long enough to be recognized

This isn't rigidity.

It's **recognizability**.

Decision-level insight

Marketing doesn't work because you're present everywhere.

It works because **everywhere you're present tells the same story**.

When signals agree:

- trust compounds
- decisions accelerate
- growth feels less fragile

When they don't:

- effort increases
- results flatten
- frustration rises

The competitive advantage going forward isn't mastering more channels.

It's building systems that **don't argue with themselves**.

Alignment is often easier to recognize externally than internally. If you want to see how your signals compare to competitors in your market, you can request a competitor alignment review at **https://patientflow.us/resources-dental** to highlight where messaging, structure, and trust signals diverge.

The next chapter explains where paid ads fit into this picture—not as a foundation, but as an accelerator that either validates alignment or exposes its absence.

CHAPTER 15

Paid Ads: Speed, Not Safety

Why this chapter matters

Paid ads are often the first thing dental practice owners turn to when growth feels urgent.

They feel concrete.

They feel controllable.

They feel fast.

And in one important way, they are.

But ads are also one of the most misunderstood parts of the digital marketing system—especially when they're expected to *replace* trust instead of **revealing whether trust already exists**.

This chapter explains where paid ads actually fit in a holistic system, why they can accelerate growth—or accelerate disappointment—and how to think about them without over-crediting or over-blaming them.

Ads solve a timing problem, not a trust problem

At their core, paid ads exist to solve one thing: **immediacy**.

They allow a business to appear:

- before organic trust has fully accumulated
- faster than Maps alone would allow
- in moments of urgency

That's valuable.

But it's also limited.

Ads don't convince people you're trustworthy.

They assume the trust question will be answered **after the click**.

If what people find after the click doesn't align with expectations, the system learns—and it charges you for the lesson.

Why ads feel easier than everything else

Ads come with dashboards.

Clear budgets.

Immediate feedback.

You turn them on.

Traffic shows up.

Something *happens*.

Compared to SEO, content, or Maps—which all feel slow and ambiguous—ads feel reassuring.

That's why so many businesses mistake **activity** for **progress**.

But ease is not the same as durability.

Ads borrow credibility from everything else

Here's the systems-level truth most people miss:

> Ads don't stand alone.
>
> They borrow credibility from your entire digital footprint.

When someone clicks an ad, they don't stop evaluating.

They start validating.

They check:

- your website structure
- your service clarity
- your reviews
- your Maps presence
- sometimes your social

If those signals align, ads feel effective.

If they don't, ads feel "expensive" or "low quality."

The ad didn't fail.

The system just exposed a weak foundation.

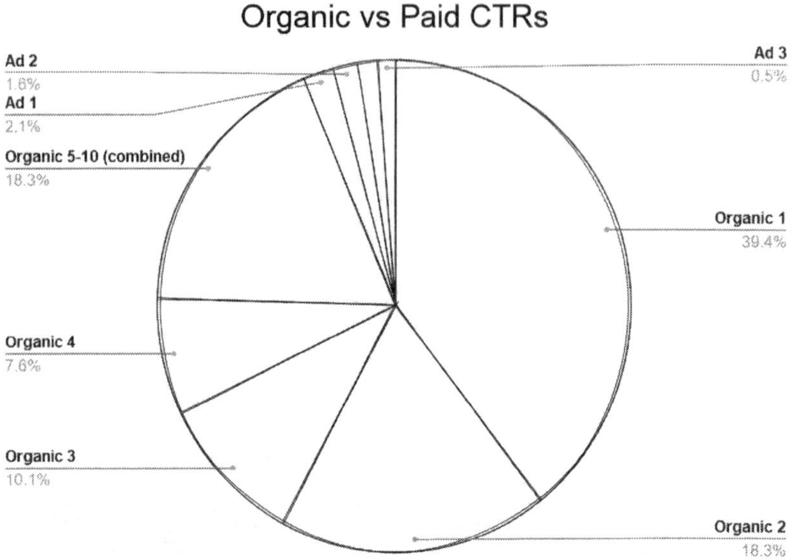

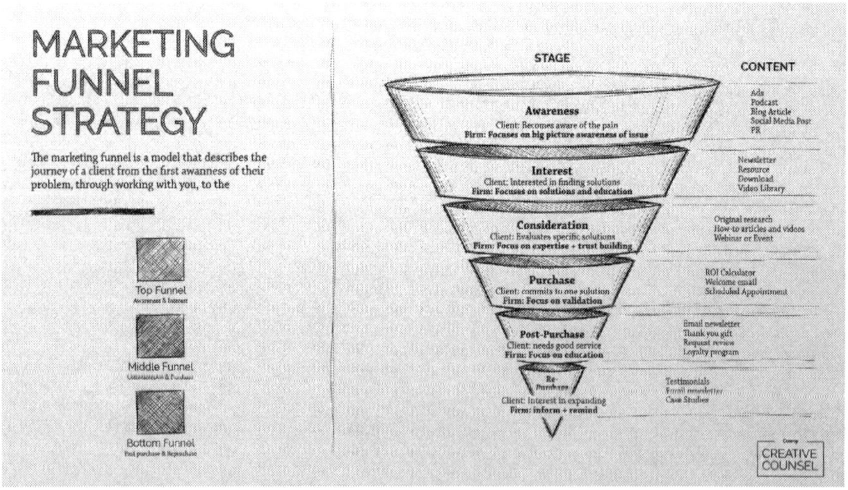

Why ads get blamed for structural problems

This pattern shows up constantly:

- Ads are launched to "kickstart growth"
- Performance is inconsistent
- Costs rise
- Conversion rates flatten
- Ads get blamed

In reality, ads often did exactly what they're designed to do:

> They surfaced demand *before* the business was ready to satisfy it efficiently.

Ads don't fix:

- unclear positioning
- vague service pages
- weak Maps presence
- inconsistent messaging

They highlight them.

Why people scroll past ads (and why that matters)

Many users—especially in trust-sensitive categories—have learned patterns.

Over time, they:

- recognize generic ad language
- associate ads with aggressive selling
- prefer "earned" placements when stakes feel personal

This doesn't mean ads don't work.

It means **ads don't get the benefit of the doubt**.

Maps and organic listings often do—because they feel selected, not purchased.

That conditioning explains why:

- ads may get impressions but fewer clicks
- Maps absorbs disproportionate attention
- organic results feel "safer" to many users

Ads as a diagnostic tool

Here's a more useful way to think about paid ads:

> Ads are a stress test for your alignment.

When ads work well:

- your messaging is clear
- your offer matches intent
- your trust signals are coherent

When ads struggle:

- the system isn't convinced

- users hesitate after the click
- something downstream is misaligned

Instead of asking:

> "How do we optimize the ads?"

The better first question is:

> "What are the ads revealing about everything else?"

Where ads *do* shine

Used correctly, ads are powerful.

They are especially useful for:

- accelerating visibility once trust exists
- supporting launches or promotions
- capturing demand during peak moments
- testing messaging *within* an aligned system

Ads work best when they:

- reinforce what organic already suggests
- echo the same language as your website
- point to pages that already convert organically

In that role, they compound momentum instead of substituting for it.

The risk of over-reliance

When ads become the *primary* growth engine, fragility increases.

Why?

Because ads are:

- rented attention
- sensitive to competition
- vulnerable to platform changes

Trust-based visibility—Maps, organic, consistent signals—compounds.

Ads reset every day.

That doesn't make ads bad.

It makes them **temporary by design**.

Decision-level insight

Paid ads are not a foundation.

They are an **accelerator**.

They amplify whatever system they're attached to:

- aligned systems scale faster
- fragmented systems burn money faster

If ads feel like the only thing holding growth together, that's not success.

It's a warning.

Used in the right place, ads provide speed without sacrificing stability.

Used as a substitute for trust, they expose problems you can't bid your way out of.

The next chapter tackles one of the biggest reasons ads, SEO, and social get misinterpreted in the first place: **measurement**, and why the numbers we're given often create more confidence than clarity.

CHAPTER 16

Measurement Without Illusions

Why this chapter matters

Measurement is where confident decisions quietly turn into bad ones.

Not because data is wrong—but because it's **misinterpreted**.

Most dental practice owners are surrounded by numbers:

- clicks
- impressions
- leads
- conversions
- dashboards full of charts

And yet, many still feel unsure whether marketing is *actually* working.

This chapter exists to explain why that confusion is so common, why modern marketing data creates **false certainty**, and how to think about measurement in a way that supports judgment instead of undermining it.

Marketing data is persuasive by design

Every major marketing platform has an incentive to make its data feel:

- precise
- attributable
- controllable

That's not malicious.

It's structural.

Platforms need you to believe:

> "This channel is responsible for this result."

So they build reporting systems that **assign credit**, even when reality is more complex.

The problem is that patient acquisition rarely follows a single, clean path.

Why attribution breaks down in real life

A typical dental practice patient journey might look like this:

- Sees your business in Maps
- Clicks an organic result days later
- Reads reviews

- Visits your website
- Follows you on social
- Sees a retargeting ad
- Finally books after a reminder

Which channel "caused" the conversion?

The honest answer is:

All of them—and none of them alone.

Attribution models don't capture reality.

They **simplify it**.

That simplification is useful—but dangerous if taken literally.

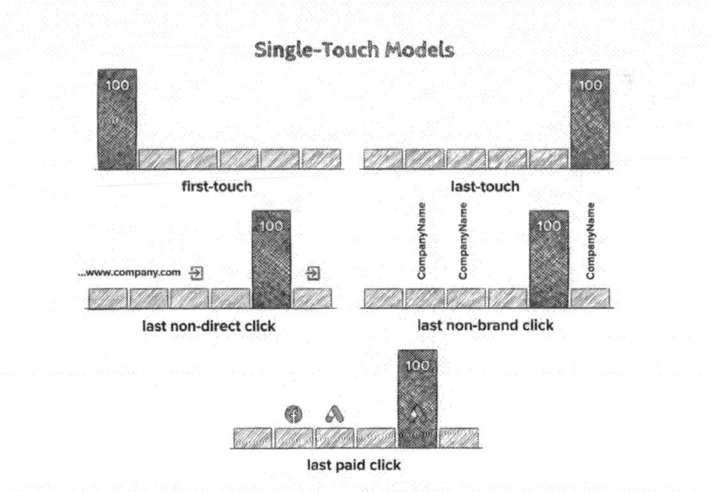

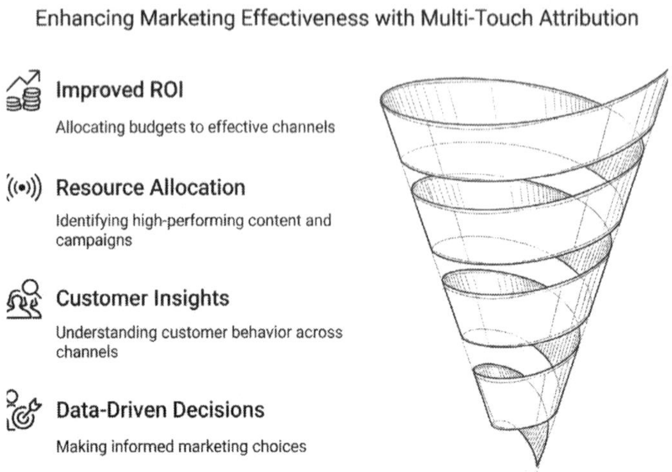

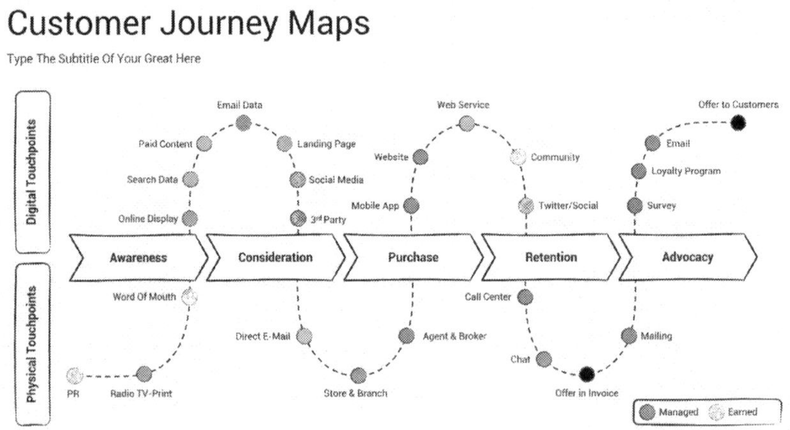

Last-click attribution is convenient—and misleading

Most reporting defaults to last-click attribution.

Whatever someone touched *most recently* gets the credit.

That makes reports easy to read.

It also quietly distorts decision-making.

Last-click attribution tends to:

- over-credit ads
- under-credit organic and Maps
- ignore trust-building steps entirely

Which leads to predictable mistakes:

- cutting long-term investments too early
- over-scaling ads that depend on organic trust
- chasing channels that "look good" on paper

The data isn't lying.

It's just incomplete.

Why dashboards create false confidence

Dashboards feel objective.

They have:

- numbers
- charts
- percentages
- trends

But most dashboards answer a narrow question:

> "What happened *inside this platform*?"

They rarely answer:

- Why did this work *now*?
- Would it work without other channels?
- Is this compounding—or temporary?

When owners rely solely on dashboards, they often confuse **visibility of data** with **clarity of understanding**.

Those are not the same thing.

What can be measured reliably—and what can't

This distinction matters.

Some things *can* be measured reasonably well:

- paid ad performance within a campaign
- conversion rates on specific pages
- call volume trends over time

Other things cannot be isolated cleanly:

- brand trust
- Maps influence on organic clicks
- social media's role in validation
- AI-driven discovery effects

Trying to force precise measurement where precision isn't possible doesn't improve insight—it creates overconfidence.

Measurement should support decisions, not justify them

Here's a subtle but important shift:

Bad measurement is used to **defend choices already made**.

Good measurement is used to **inform the next decision**.

That means asking better questions, such as:

- Are outcomes becoming more consistent over time?
- Is dependence on any one channel decreasing?
- Are ads becoming more efficient as organic improves?
- Is visibility expanding *without* proportional spend increases?

These questions don't live in a single dashboard.

They live across systems.

Why holistic systems are harder to measure—but more reliable

Aligned marketing systems are paradoxical.

They:

- produce better outcomes
- feel harder to attribute
- look "messier" in reports

That's because success emerges from **reinforcement**, not isolation.

When everything supports everything else:

- no single channel deserves full credit
- no single metric explains success
- results feel steadier, not spiky

This frustrates people looking for simple answers.

It rewards people making long-term decisions.

How AI complicates measurement further

AI introduces more intermediated behavior:

- synthesized answers
- blended discovery paths
- fewer obvious clicks

Which means:

- attribution becomes fuzzier, not clearer
- surface-level metrics matter less
- system-wide outcomes matter more

Trying to measure AI-driven discovery with pre-AI tools is like timing traffic with a sundial.

The tools lag the reality.

Decision-level insight

Measurement is not about knowing *exactly* what worked.

It's about knowing **enough to choose wisely next**.

The most reliable signal of healthy marketing isn't perfect attribution.

It's this:

- growth becomes less fragile
- fewer emergencies drive decisions
- channels reinforce instead of compensate

When marketing systems are aligned, results feel calmer—even when measurement feels less precise.

When dashboards stop answering useful questions, directional reviews can be more clarifying than precise metrics. Some owners step back and review overall system consistency using structured evaluations at **https://patientflow.us/resources-dental**, rather than chasing attribution certainty.

The next chapter looks at why AI magnifies this dynamic, and why it acts as a force multiplier for coherence—or a landmine for fragmentation.

CHAPTER 17

What AI Is Actually Changing: Pattern Recognition at Scale

Why this chapter matters

By the time most dental practice owners notice AI, they're already anxious.

They hear:

- "Search is changing."
- "SEO is dying."
- "AI will replace websites."
- "You need to pivot—now."

Most of that noise comes from misunderstanding *what* AI actually changes—and what it doesn't.

This chapter exists to calm the conversation by being precise.

AI is not rewriting the rules of marketing.

It is **enforcing them faster and more consistently**.

And that distinction matters, because it determines whether AI becomes a force multiplier—or a landmine.

AI doesn't create intent—it interprets it

Let's start with what *hasn't* changed.

People still:

- want reassurance before making decisions
- compare options
- seek proximity, legitimacy, and competence
- hesitate when signals conflict

AI didn't invent those behaviors.

It learned them.

What AI changes is **how efficiently systems interpret patterns** that already exist.

Instead of evaluating signals slowly and partially, AI can:

- synthesize across surfaces
- detect repetition and contradiction
- summarize themes at scale

That doesn't alter human psychology.

It accelerates how platforms respond to it.

From ranking pages to summarizing reality

Traditional search focused heavily on:

- pages
- links
- positions

AI shifts the emphasis toward:

- entities
- themes
- agreement across sources

Instead of asking:

> "Which page should rank here?"

AI-assisted systems increasingly ask:

> "What seems to be true about this business, based on everything we can observe?"

That's a subtle but profound shift.

It favors:

- businesses with clear identities
- consistent messaging
- stable service offerings

And it exposes:

- fragmented brands
- contradictory signals
- content created without structure

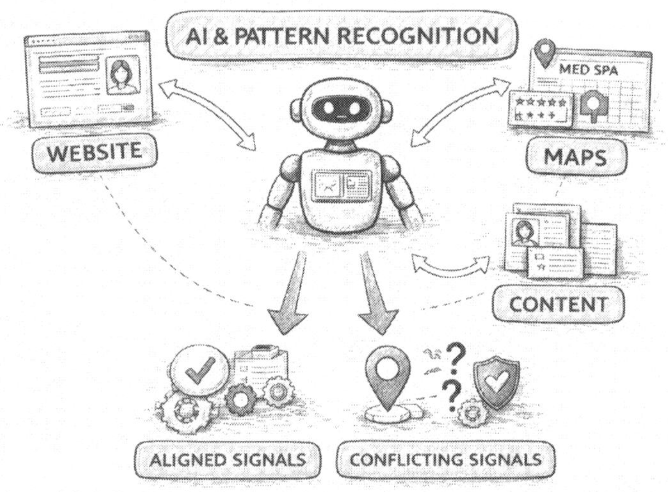

Why coherence becomes more valuable than optimization

In a pre-AI world, it was sometimes possible to:

- optimize one page
- exploit one gap
- lean heavily on one channel

AI makes that harder.

When systems summarize across:

- website structure
- Maps data
- reviews
- content
- social presence

They don't just see *what you optimized*.

They see **what you consistently represent**.

That's why AI rewards coherence more than cleverness.

Clever tactics spike.

Coherent systems compound.

AI widens gaps—it doesn't level the field

There's a comforting myth that AI "democratizes" marketing.

In practice, it does the opposite.

AI:

- helps strong systems scale faster
- helps weak systems unravel faster
- reduces the margin for ambiguity

Businesses with:

- clear positioning
- aligned signals

- stable messaging

become easier to understand—and therefore easier to recommend.

Businesses with:

- scattered content
- conflicting channels
- reactive strategies

become harder to summarize—and therefore safer to avoid.

The gap widens quietly, not dramatically.

Why this feels destabilizing right now

We're in a transition period.

AI systems are:

- learning what to trust
- testing how to present summaries
- adjusting how much autonomy they exercise

That creates volatility.

But volatility doesn't mean chaos.

It means **re-weighting**.

Signals that used to be marginal are becoming decisive.

Signals that used to be isolated are now compared.

This is uncomfortable—but predictable.

What AI is *not* doing

It's important to be clear about what AI is *not* replacing:

- It's not replacing human judgment in decisions
- It's not eliminating trust evaluation
- It's not bypassing the need for legitimacy
- It's not rewarding volume for its own sake

AI doesn't remove the need for structure.

It **punishes the absence of it**.

Why dental practices are especially exposed

Dental practices sit at a sensitive intersection:

- health-adjacent
- Health outcomes
- locally constrained
- trust-dependent

That makes AI systems more cautious, not more permissive.

The more personal the outcome, the more AI leans on:

- corroboration
- repetition

- stability

Which means dental practices don't get to be vague.

They get to be clear—or invisible.

Decision-level insight

AI doesn't ask whether your marketing is creative.

It asks whether it is **internally consistent**.

If your:

- website says one thing,
- Maps suggests another,
- content drifts,
- social contradicts,
- ads promise something else,

AI doesn't debate which is correct.

It simply hesitates.

But when signals agree, AI becomes an amplifier.

The next chapter explains how AI affects *execution*—how it changes the speed, scale, and risk of marketing work itself—and why aligned systems benefit first while chaotic ones break faster.

CHAPTER 18

AI in Marketing: Speed, Scale, and Signal Amplification

Why this chapter matters

By now, it should be clear that AI doesn't change *what* marketing rewards.

It changes **how fast those rewards—and penalties—show up**.

This chapter exists to explain how AI affects the *execution* of marketing work itself: content creation, analysis, iteration, and optimization. Not at a tactical level, but at a systems level—where decisions compound or collapse.

The danger with AI isn't misuse.

It's **mis-sequencing**.

Used inside a coherent system, AI accelerates momentum.

Used inside a fragmented one, it amplifies confusion at machine speed.

AI is a force multiplier, not a strategist

AI is very good at:

- producing drafts
- summarizing information
- identifying patterns
- accelerating repetitive tasks

AI is not good at:

- deciding what matters
- choosing priorities
- understanding local nuance
- judging risk

That distinction is critical.

AI doesn't create strategy.

It **executes whatever strategy already exists**—whether that strategy is sound or not.

This is why two dental practices can use the same AI tools and see opposite outcomes.

Speed changes the cost of mistakes

Before AI, poor marketing decisions were often self-limiting.

Content took time.

Iteration was slow.

Bad ideas burned out naturally.

AI removes that friction.

Now:

- bad content can be produced in volume
- misaligned messaging can spread everywhere
- weak structure can be reinforced repeatedly

The cost of being wrong hasn't disappeared.

It's **front-loaded**.

Mistakes surface faster.

Damage compounds sooner.

Why AI punishes inconsistency first

AI systems don't evaluate content the way humans do.

They don't ask:

> "Is this persuasive?"

They ask:

> "Is this consistent with everything else we see?"

Which means:

- repeated themes strengthen signals
- small contradictions stand out more clearly

- drift becomes obvious quickly

A human might forgive inconsistency.

AI flags it.

That's why AI adoption often *reveals* structural problems before it creates benefits.

Where AI actually helps aligned dental practices

Inside a coherent system, AI is genuinely powerful.

It can:

- accelerate content production *within defined themes*
- support research and synthesis
- maintain tone and terminology consistency
- assist with internal documentation and planning

Most importantly, AI reduces the **cost of repetition**.

And repetition—when aligned—is what trust systems reward.

Why "more content" is the wrong AI goal

One of the most dangerous assumptions is:

> "If AI can create content, we should create a lot of it."

Volume without structure doesn't scale trust.

It scales noise.

AI should be used to:

- reinforce existing service clusters
- deepen explanations that already matter
- support content that already converts organically

AI is best at saying the **same important things more clearly, more often, and in more formats**.

Not at inventing new directions.

AI makes silos visible—and expensive

Before AI, different teams or vendors could operate in isolation.

The website said one thing.

Social did another.

Ads promised something else.

Blogs wandered.

AI doesn't tolerate that separation.

When AI synthesizes across channels, silos collide.

That collision shows up as:

- unclear summaries
- weaker recommendations
- conservative visibility

What used to be invisible becomes obvious.

Human judgment becomes more valuable, not less

As execution speeds up, judgment becomes the bottleneck.

Someone still has to decide:

- which services matter most
- which questions are worth answering
- which signals should be reinforced
- which risks aren't worth taking

AI doesn't eliminate that responsibility.

It **raises the stakes** of getting it right.

This is why the most successful AI adoption looks boring:

- careful
- constrained
- focused
- repetitive

Boring scales.

Chaotic breaks.

The quiet risk of AI "efficiency"

Efficiency sounds positive.

But efficiency without direction is dangerous.

AI makes it easier to:

- do the wrong thing faster
- justify volume over clarity
- mistake activity for progress

The question isn't:

> "Can we do this with AI?"

It's:

> "Should this exist at all?"

AI doesn't answer that.

You do.

Decision-level insight

AI doesn't reward creativity, speed, or output.

It rewards **coherence at scale**.

If your marketing system:

- already agrees with itself
- already reinforces the same truths

- already builds trust consistently

AI becomes a powerful ally.

If it doesn't, AI becomes a spotlight—revealing contradictions you could once hide behind effort.

The next chapter addresses the boundary where speed, scale, and trust collide most sharply for dental practices: **compliance**, and why treating it as a system constraint—not a legal footnote—is essential in an AI-accelerated world.

CHAPTER 19

Compliance as a Marketing Constraint (Not a Legal Footnote)

Why this chapter matters

Most dental practice owners think about compliance only when something goes wrong.

A warning.

A complaint.

A platform rejection.

Until then, compliance often lives in a separate mental box—handled by legal counsel, staff training, or policy documents, but rarely integrated into marketing strategy itself.

That separation no longer works.

In an AI-accelerated, trust-sensitive category like dental practices, **compliance is not an external rule set**. It's a **design constraint** that shapes what the system is willing to trust, amplify, and recommend.

This chapter exists to reframe compliance—not as friction, but as structure.

Compliance isn't about rules—it's about risk signaling

From a platform's perspective, compliance is not primarily about legality.

It's about **risk containment**.

Search engines, ad platforms, social networks, and AI systems all share a common incentive:

> Avoid distributing content that could plausibly lead to harm, regret, or liability.

In healthcare-adjacent categories, that threshold is lower.

The margin for error is thinner.

And the tolerance for ambiguity is smaller.

Compliance signals help systems decide:

- how cautious to be
- how much reach to allow
- how confidently to surface a business

Why dental practices sit in a uniquely sensitive zone

Dental practices are not hospitals.

But they are not consumer lifestyle businesses either.

They operate in a middle ground where:

- outcomes affect long-term health and quality of life
- decisions feel consequential, not casual
- expectations must be set and managed carefully

That positioning triggers higher scrutiny—especially as AI systems become better at classifying clinical risk and patient consequence.

This doesn't put dental practices at a disadvantage.

It means clarity, consistency, and restraint matter more than ever.

What compliance looks like to machines

Humans think of compliance in terms of:

- consent forms
- disclosures
- internal policies

Machines infer compliance from **patterns**.

They look for:

- exaggerated claims vs measured language
- consistency between promises and explanations
- alignment between visuals, copy, and outcomes
- absence of sensationalism

When marketing feels careful, specific, and grounded, systems interpret that as **lower risk**.

When marketing feels aggressive, vague, or inflated, systems pull back—not because they're moral, but because they're defensive.

Why AI raises the stakes

Before AI, inconsistent or risky messaging could slip through unnoticed.

AI changes that by:

- comparing claims across pages
- summarizing themes across channels
- identifying contradictions at scale

This means:

- overpromising gets flagged faster
- vague language stands out more

- mismatches between ads, website, and social are easier to detect

AI doesn't need to "understand" HIPAA to enforce caution.

It only needs to recognize **patterns associated with risk**.

Compliance supports trust—when it's integrated

Here's the important shift:

Compliance works best when it's **embedded**, not appended.

When:

- service pages explain limits as clearly as benefits
- blog content educates without sensationalizing
- social posts reflect professionalism, not hype
- ads promise exactly what the site supports

Compliance stops feeling restrictive.

It starts functioning as **credibility reinforcement**.

The system sees:

> "This business understands the responsibility that comes with its category."

That understanding is rewarded.

The danger of treating compliance as a checkbox

When compliance is treated as an afterthought:

- marketing pushes boundaries
- disclaimers feel bolted on
- tone becomes inconsistent

That creates mixed signals.

And mixed signals are exactly what modern systems avoid amplifying.

This is especially risky with AI-assisted content, where speed can outpace judgment if constraints aren't clear upfront.

Why restraint scales better than boldness

In many industries, bold claims drive attention.

In dental practices, bold claims raise red flags.

Restraint:

- signals professionalism
- reduces perceived risk
- increases long-term stability

This doesn't mean being dull or invisible.

It means being **precise instead of promotional**.

Precision is safer.

And safety is rewarded with reach.

Compliance as competitive advantage

Here's the counterintuitive insight:

As AI increases scrutiny, businesses that are:

- clear about what they do
- honest about outcomes
- consistent about messaging

gain advantage over those that chase attention.

Not because they're louder—but because they're easier to trust.

Compliance isn't a brake.

It's a **guardrail** that keeps your marketing system moving forward without veering into territory that stalls growth later.

Decision-level insight

Compliance isn't something you add *after* marketing is built.

It's something you design **into the system from the beginning**.

In an AI-driven landscape, trust-sensitive categories don't win by pushing limits.

They win by being:

- coherent
- measured
- aligned

- predictable in the best way

When compliance supports your structure instead of fighting it, marketing becomes safer, steadier, and more scalable.

The next chapter moves from external visibility to internal reality—how AI affects **operations and patient experience**, and why supporting staff (not replacing them) is the most durable path forward.

CHAPTER 20

AI in Operations and Patient Experience

Why this chapter matters

Up to now, we've talked about AI primarily as a *marketing* force—how it affects visibility, trust signals, and discovery.

But for dental practices, some of AI's most consequential impact happens **after** marketing has done its job.

This chapter exists to explain where AI fits inside operations and patient experience—not as a replacement for people, but as a **support system** that either strengthens trust or quietly erodes it.

In trust-sensitive businesses, how patients are handled matters just as much as how they're acquired.

Patient experience is part of the marketing system

Here's a subtle but critical truth:

> From the system's perspective, marketing doesn't end at the click or the call.

Platforms infer quality not just from:

- what you say
- how visible you are

but from:

- whether people follow through
- whether interactions feel smooth
- whether outcomes align with expectations

Operations feed back into marketing—whether you intend them to or not.

Missed calls, confusing booking flows, delayed follow-ups, and inconsistent communication all show up indirectly as weaker trust signals.

Where AI actually helps operations (when used carefully)

AI excels at **supporting repetitive, low-judgment tasks**.

In dental practice operations, that often includes:

- call routing and triage
- appointment reminders
- intake organization
- post-visit follow-up
- basic FAQ handling

Used correctly, AI:

- reduces friction

- improves responsiveness
- supports staff under load

Importantly, it does this **without changing the core human relationship**.

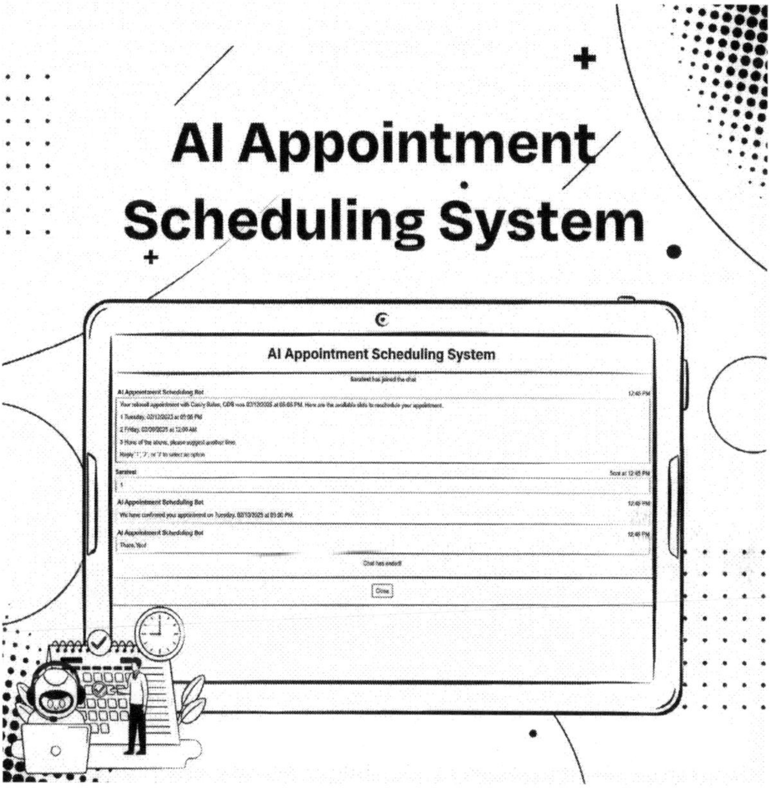

Why replacing humans is the wrong instinct

AI often gets pitched as a way to "reduce staff costs" or "eliminate bottlenecks."

That framing is dangerous in dental practices.

Patients don't just want efficiency.

They want reassurance.

Replacing human interaction too aggressively:

- increases perceived risk
- reduces emotional confidence
- introduces uncertainty at the exact wrong moment

AI should absorb *administrative load*, not replace *human judgment*.

The goal is to make staff better—not invisible.

Consistency matters more than sophistication

From a systems perspective, consistency beats cleverness.

AI-driven communication should:

- sound like the rest of your brand
- align with your website language
- reflect the same expectations your marketing sets

A simple, consistent reminder message builds more trust than a sophisticated but mismatched one.

Inconsistency creates cognitive friction.

And friction shows up as hesitation.

Why AI can quietly damage trust if misused

Poorly implemented AI often fails in subtle ways:

- answers too confidently
- provides incomplete context
- feels generic or evasive
- escalates issues poorly

None of these trigger immediate alarms.

They just reduce confidence.

Over time, that reduction shows up as:

- fewer completed bookings
- more no-shows
- more follow-up questions
- lower conversion from inquiry to visit

The system interprets those outcomes—not the intent behind the tool.

AI adoption curves are human, not technical

Most resistance to AI inside dental practices isn't about fear of technology.

It's about:

- loss of control

- fear of replacement
- concern over patient trust

Ignoring that resistance creates internal friction that spills outward.

Successful AI adoption:

- involves staff early
- clarifies boundaries
- positions AI as support
- evolves gradually

Adoption speed matters less than **adoption alignment**.

HIPAA awareness as an operational signal

Just as compliance shapes marketing trust, it shapes operational trust too.

AI systems that:

- handle sensitive data carelessly
- blur boundaries
- feel opaque

introduce risk the system notices.

Careful implementation—clear roles, limited scope, defined escalation paths—signals responsibility.

Again, this isn't about perfection.

It's about **predictability**.

The feedback loop most businesses miss

Operations influence:

- reviews
- engagement patterns
- repeat visits
- brand perception

Those influence:

- Maps visibility
- organic trust
- ad performance

Which feeds back into:

- acquisition cost
- growth stability

AI sits *inside* that loop.

When it supports clarity and consistency, everything strengthens.

When it introduces confusion, everything softens.

Decision-level insight

AI in operations should never be about replacing people.

It should be about **protecting the patient experience from avoidable friction**.

In dental practices, trust is built through:

- smooth handoffs
- clear communication
- human reassurance at key moments

AI helps when it reinforces those moments.

It hurts when it shortcuts them.

With AI integrated thoughtfully, operations become an extension of marketing—not a contradiction of it.

The final section of the book now turns fully toward **owner decisions**—how to hire, outsource, and structure marketing in a way that respects everything you've just learned, without chasing tools, vendors, or false certainty.

CHAPTER 21

How to Hire (Or Fire) a Marketing Company

Why this chapter matters

Most dental practice owners don't fail at marketing because they chose the *wrong* channel.

They fail because they delegated judgment.

Hiring a marketing company feels like a solution to complexity. And in many cases, it can be—*if* the relationship is built around the right expectations. But when hiring becomes a way to avoid understanding how the system works, it often creates more confusion, not less.

This chapter exists to help you evaluate marketing partners **without becoming a marketer**, and to recognize when a relationship is structurally misaligned—even if the vendor is competent, busy, and well-intentioned.

The uncomfortable truth about outsourcing marketing

Here's the reality most owners eventually discover:

> You can outsource execution.
>
> You cannot outsource responsibility.

No agency—no matter how experienced—can:

- decide your risk tolerance
- choose which tradeoffs matter
- understand your internal constraints
- feel the cost of misalignment the way you do

When owners abdicate those decisions, marketing becomes reactive, fragmented, and easy to overspend on.

Good vendors don't remove complexity.

They **require informed leadership to work well**.

Why most vendor conversations start in the wrong place

Vendor conversations usually begin with:

- deliverables
- pricing
- timelines
- channels

Those are easy to talk about.

They are also the least important things to discuss first.

The more important questions are:

- How does this company think about trust?
- How do they explain cause and effect?
- Do they understand dental practices as a category—or just as keywords?
- Can they explain *why* something should work, not just *what* they'll do?

If a vendor can't explain systems clearly, they won't manage them well.

What good answers sound like

You don't need technical depth.

You need **conceptual clarity**.

Strong partners tend to:

- speak in terms of alignment, not hacks
- explain tradeoffs openly
- set expectations conservatively
- distinguish between short-term visibility and long-term trust
- acknowledge uncertainty without hand-waving

They are comfortable saying:

> "This depends."
>
> "This will take time."
>
> "This only works if these other pieces are in place."

That's not hesitation.

That's honesty.

Red flags that matter more than tactics

Some warning signs have nothing to do with skill level.

Be cautious when a vendor:

- promises outcomes without discussing competition
- avoids explaining how Maps, organic, ads, and content interact
- focuses on isolated metrics instead of system behavior
- dismisses concerns about compliance or risk
- treats AI as a shortcut rather than an amplifier

These are not sales issues.

They're **incentive mismatches**.

Why “results-based” language is often misleading

Owners understandably want accountability.

But beware of guarantees framed around:

- rankings
- traffic volume
- lead counts
- platform-specific metrics

These are easy to report—and easy to game.

What’s harder (and more meaningful) is:

- improved consistency across channels
- reduced dependence on paid traffic
- stronger performance from the same spend
- fewer emergency decisions
- provider comparison
- patient hesitation

If a vendor can’t talk about those outcomes, they may be optimizing for optics, not durability.

Timelines: the most common source of disappointment

Many marketing relationships sour not because nothing works—but because **expectations were mismatched**.

Trust-based systems move slowly at first.

Then they compound.

A good partner will:

- explain where progress shows up early (often invisibly)
- clarify when visible movement becomes likely
- distinguish between effort and signal accumulation

If everything is framed as "quick wins," you're being sold speed—not stability.

When it's time to fire a vendor

This is difficult—but sometimes necessary.

It's usually time to part ways when:

- explanations stop making sense
- questions are deflected, not answered
- strategy changes constantly without clear reasons
- activity increases but clarity decreases
- alignment across channels worsens over time

Firing isn't about blame.

It's about **resetting responsibility**.

What the best owner–vendor relationships look like

The healthiest relationships feel less like outsourcing and more like collaboration.

Owners:

- understand the system at a high level
- make prioritization decisions
- hold coherence as a non-negotiable

Vendors:

- execute within those constraints
- surface tradeoffs early
- adjust tactics without shifting strategy
- measure progress holistically

No one pretends marketing is simple.

But no one treats it as mystical either.

Decision-level insight

Hiring a marketing company doesn't eliminate the need for judgment.

It **raises the value of it**.

The best partners don't ask you to trust them blindly.

They help you see the system clearly enough to make informed decisions—then execute within that clarity.

If a vendor makes marketing feel more confusing over time, something is wrong.

If they make it feel more legible—even when progress is slow—you're likely on the right path.

If you're preparing to interview agencies—or reevaluate an existing relationship—it can help to walk in with better questions. Some readers use vendor interview preparation frameworks at **https://patientflow.us/resources-dental** to clarify what they're actually listening for during those conversations.

The next chapter tackles a related decision many owners face: whether to build internally, outsource fully, or adopt a hybrid approach that balances control with leverage.

CHAPTER 22

DIY vs. Outsource: An Honest Framework

Why this chapter matters

Once owners understand how digital marketing actually works, a new tension appears.

It's no longer:

> "Should I do marketing?"

It becomes:

> "How much of this should I own—and how much should I delegate?"

This is one of the most consequential decisions in the entire system, because it determines not just cost, but **control, coherence, and long-term resilience**.

This chapter exists to give you a clear, judgment-based framework—without pretending there's a single correct answer.

The real tradeoff isn't money—it's judgment

DIY vs. outsource is usually framed as:

- cost vs. expertise
- time vs. money

That framing misses the real issue.

The true tradeoff is:

> **Who holds the mental model of the system?**

If no one inside your business understands how marketing works:

- vendors drift
- channels fragment
- decisions become reactive

If someone *does* understand the system—even imperfectly:

- outsourcing becomes safer
- vendors perform better
- alignment is easier to maintain

DIY and outsourcing are not opposites.

They're **leverage choices**.

What DIY actually means (and doesn't)

DIY does not mean:

- building everything yourself forever
- becoming a technical specialist
- running ads, SEO, social, and AI personally

DIY means:

- owning the **thinking layer**
- understanding cause and effect
- deciding priorities and boundaries

Execution can still be delegated.

Owners who confuse DIY with execution often burn out.

Owners who confuse outsourcing with abdication often stall.

When DIY makes sense

DIY-heavy approaches tend to work best when:

- budgets are constrained
- timelines are flexible
- owners are willing to learn at a conceptual level
- the business is early or re-centering

DIY forces clarity.

It exposes gaps.

It makes tradeoffs visible.

That learning has long-term value—even if execution is later outsourced.

When full outsourcing makes sense

Outsourcing-heavy approaches tend to work best when:

- the owner already understands the system
- internal priorities are clear
- expectations are realistic
- alignment is actively managed

Outsourcing works when it's guided.

It fails when it's blind.

The problem isn't vendors.

It's unclear leadership.

The hybrid model (where most durable systems land)

For most dental practices, the most resilient structure is hybrid.

That usually looks like:

- **Internal ownership of strategy and coherence**
- **External execution of specialized tasks**

Examples:

- Internal clarity on services, positioning, tone, and risk
- External support for SEO execution, ad management, content production, or tooling
- Internal oversight that ensures everything agrees

This model scales without surrendering judgment.

Why internal alignment matters more than internal labor

It's tempting to equate "in-house" with "control."

But control doesn't come from who does the work.

It comes from **who defines the rules**.

A small internal team with clarity outperforms:

- a large agency with vague direction
- multiple vendors working in isolation
- reactive tool adoption

Alignment beats manpower.

The cost of partial understanding

One of the most dangerous states is *half-knowledge*.

Owners know just enough to:

- recognize problems
- distrust vendors

- feel overwhelmed

But not enough to:

- diagnose causes
- set constraints
- make confident decisions

This is where marketing becomes exhausting.

The solution isn't more tactics.

It's deeper **conceptual understanding**—exactly what this book is meant to provide.

AI makes this decision more important, not less

AI lowers execution cost.

It does not lower the cost of bad judgment.

Which means:

- outsourcing without understanding is riskier
- DIY without constraints is faster—but more dangerous
- hybrid models benefit the most

AI doesn't remove the need to decide.

It **amplifies the consequences of decisions already made**.

Decision-level insight

You don't need to do everything yourself.

But you do need to **understand what you're delegating**.

The healthiest marketing systems:

- keep judgment close to ownership
- push execution outward
- maintain alignment as a non-negotiable

DIY vs. outsource isn't a one-time choice.

It's an evolving balance.

The final chapters bring everything together—focusing on what actually scales over time, and why dental practices that understand systems are still early in a market that's about to widen dramatically.

CHAPTER 23

What Actually Scales

Why this chapter matters

By this point, you've seen enough to recognize a pattern.

The dental practices that grow steadily over time aren't necessarily:

- the most aggressive,
- the most creative,
- or the most active on every channel.

They're the ones that **compound trust without constantly reintroducing confusion**.

This chapter exists to clarify what *actually* scales in digital marketing—and, just as importantly, what only *looks* like scale for a while.

Scale is not growth—it's repeatability

Growth can happen accidentally.

Scale cannot.

Scale requires that something works:

- more than once,
- under different conditions,
- without constant heroics.

In digital marketing, that means your system has to function even when:

- ads are paused,
- algorithms shift,
- staff changes,
- platforms introduce new features.

If results depend on constant intervention, they're not scalable.

They're fragile.

Why hacks don't compound

Shortcuts feel attractive because they promise leverage.

And sometimes they work—briefly.

But hacks fail to scale because:

- they exploit gaps instead of building foundations,

- they require constant novelty,
- they break when the system adapts.

Trust-based systems, by contrast:

- move slowly at first,
- feel boring while they're building,
- become hard to displace once established.

This is why "clever" strategies tend to spike and fade, while simple, consistent ones quietly win.

What actually compounds in dental practice marketing

Across everything we've discussed, a small number of things reliably compound:

- **Clarity**
  The system understands exactly what you do and who you serve.
- **Consistency**
  That understanding doesn't change week to week or channel to channel.
- **Corroboration**
  Multiple surfaces tell the same story independently.
- **Restraint**
  Claims stay within what can be defended and delivered.

None of these are exciting.

All of them scale.

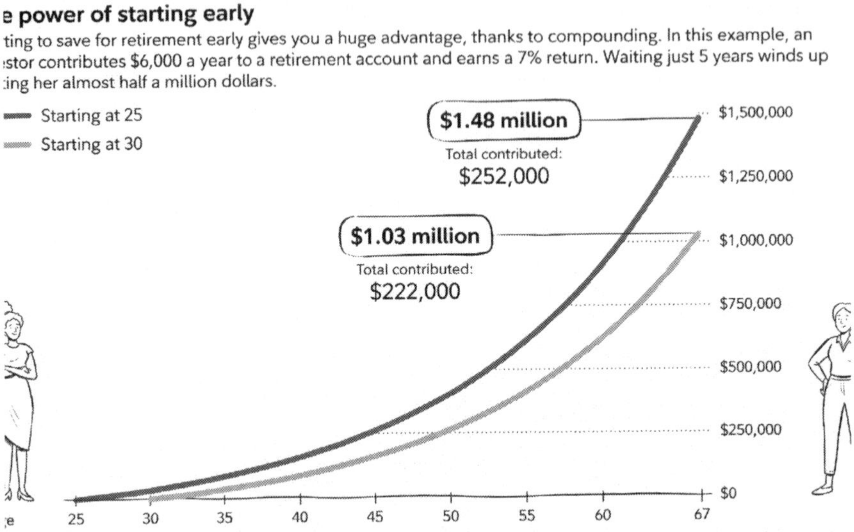

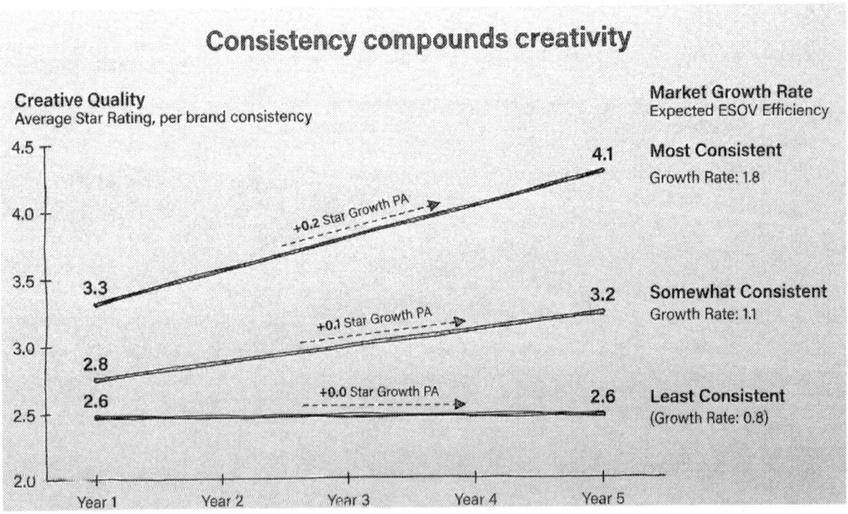

Why simplicity outperforms sophistication

As systems grow more complex, simplicity becomes an advantage.

Complex strategies:

- are harder to maintain,
- break more easily,
- depend on specific people or tools.

Simple strategies:

- survive turnover,
- adapt more easily,
- remain legible to both humans and machines.

This doesn't mean *unsophisticated*.

It means **structurally clean**.

When marketing scales, it usually gets simpler—not more elaborate.

The quiet role of time

Time is the one variable most strategies try to eliminate.

But in trust-sensitive categories like dental practices, time is not the enemy.

It's the filter.

Time:

- exposes inconsistency,
- rewards stability,
- separates signal from noise.

Systems that rely on time to work feel slow.

Systems that fight time feel urgent—and brittle.

Scale happens when you stop racing the clock and start letting alignment do its work.

Why AI favors scalable systems

AI doesn't create scale.

It removes friction.

Which means it disproportionately benefits:

- systems that are already coherent,
- messaging that is already consistent,
- structures that already make sense.

If your system scales manually, AI helps it scale faster.

If it doesn't, AI just accelerates the breakdown.

This is why AI adoption often *reveals* who was ready—and who wasn't.

What "scaling" should feel like

Healthy scale has a distinct feel:

- fewer emergencies,
- fewer pivots,
- fewer explanations,
- more predictability.

Marketing conversations shift from:

> "Why isn't this working?"

to:

> "How do we protect what's working?"

That shift is the real milestone.

Decision-level insight

What scales is not effort.

It's **alignment sustained over time**.

Dental practices that win long-term don't do more.

They contradict themselves less.

When clarity compounds, growth stops feeling fragile.

It starts feeling inevitable.

The final chapter steps back from tactics, channels, and systems to answer the question this book opened with—what kind of moment dental practice marketing is actually in, and why the opportunity is still wide open for owners who understand what you now understand.

CHAPTER 24

The Real Opportunity

Why this chapter matters

AI didn't create opportunity.

It **clarified it**.

It's making some businesses easier to find, easier to trust, and easier to choose.

And it's making others quietly disappear—not because they're bad, but because they're confusing.

This final chapter exists to name what this moment really is—and what it asks of owners.

Why dental practices are still early

Despite the noise, most dental practices are still:

- fragmented across channels,
- inconsistent in messaging,
- reactive in decision-making.

Very few operate as **coherent digital systems**.

That's not a criticism.

It's an opportunity.

The bar isn't impossibly high.

It's just misunderstood.

Owners who grasp systems—not tactics—are still in the early innings.

Why AI widens gaps instead of closing them

Every major technology shift does this.

AI:

- reduces the cost of execution,
- increases the cost of poor judgment,
- amplifies existing structure.

Businesses that:

- already make sense,
- already agree with themselves,
- already respect trust constraints,

gain disproportionate advantage.

Businesses that chase tools without clarity don't fail loudly.

They fade.

The real risk isn't change—it's misinterpretation

The danger right now isn't that marketing is changing too fast.

It's that owners:

- mistake volatility for chaos,
- mistake speed for progress,
- mistake activity for alignment.

Those mistakes lead to:

- constant vendor switching,
- tool fatigue,
- escalating spend without compounding returns.

Understanding the system doesn't freeze change.

It lets you **navigate it calmly**.

Why this moment rewards decisive owners

Decisiveness here doesn't mean rushing.

It means:

- choosing coherence over cleverness,
- structure over hacks,
- understanding over delegation.

Owners who make those choices don't need perfect execution.

They need *direction*.

Once direction is clear, execution can be improved.

Without it, execution only adds noise.

What this book was really about

This was never a book about:

- SEO tricks,
- ad tactics,
- AI tools.

It was about **legibility**.

Making marketing legible enough that:

- vendors can be evaluated,
- investments can be sequenced,
- risks can be managed,
- growth can be intentional.

When marketing becomes legible, it stops being intimidating.

It becomes governable.

Opportunity, clarified

This moment *is* a gold rush.

But not for everyone.

It's a gold rush for owners who:

- understand incentives,
- respect trust systems,
- build alignment patiently,
- use AI as a multiplier—not a shortcut.

For everyone else, it's not dramatic failure.

It's quiet irrelevance.

And the difference between those outcomes is rarely talent or budget.

It's understanding.

Final decision-level insight

Digital marketing was never mysterious.

It was just noisy.

AI didn't change the rules.

It turned the volume up.

If your system:

- agrees with itself,
- reinforces trust,
- compounds clarity,

this moment works in your favor.

Not overnight.

But reliably.

And in dental practice marketing, reliability is the rarest—and most valuable—advantage there is.

Absolutely — below is a **fully rewritten, expanded Appendix section**, incorporating *all* the recommendations we discussed:

- More **gravity**, not tactics
- Clear **consequences** when models are misunderstood
- Deeper **vendor interview framework**
- Decision-pressure framing throughout
- One reflective, human closing section
- Still calm, non-salesy, and consistent with the book's tone

This is written to be **final-manuscript quality**. You should be able to drop this in as-is.

APPENDIX

How to Think, Evaluate, and Decide

This appendix exists to do something the main chapters intentionally avoided.

The chapters explained how modern digital marketing systems behave, why platforms act the way they do, and how AI amplifies structure rather than replacing it. This appendix is about **application without reduction**—how to hold onto that understanding when real decisions, real vendors, and real pressure enter the picture.

Nothing here is a checklist.

Nothing here is mandatory.

Think of this section as a place to return when marketing starts to feel louder than it should.

APPENDIX A

The Mental Models That Govern Dental Marketing

Most marketing advice fails not because it's wrong, but because it's incomplete. It explains *what* to do without explaining *what breaks when the underlying assumptions are wrong*.

The mental models below are the quiet rules the system follows —whether you acknowledge them or not.

1. Marketing as a Trust Inference System

Platforms do not reward effort.

They infer trust.

Every signal your business sends—Maps data, website structure, content themes, reviews, social presence, ads, and even operational follow-through—feeds a single, ongoing question:

> "How confident are we that recommending this business will not lead to regret?"

This inference is probabilistic, not moral. Platforms are not judging you. They are managing risk at scale.

What breaks when this is misunderstood

When owners believe marketing rewards effort, they tend to:

- produce more content instead of clearer content
- add channels instead of resolving contradictions
- tolerate misalignment because "something is happening"

The system doesn't punish this behavior loudly.

It simply hesitates.

And hesitation looks like stalled visibility, volatile performance, and rising costs without obvious cause.

This is why many businesses feel busiest right before growth slows.

This matters most when...

You're doing more than ever, but confidence isn't increasing—yours or the system's.

2. Alignment Beats Optimization

Optimization improves parts.

Alignment improves outcomes.

A perfectly optimized channel inside a contradictory system does not compound. It compensates—and compensation has diminishing returns.

Alignment means that Maps, website, content, social, ads, and operations all reinforce the same understanding of who you are and what you offer.

What breaks when this is misunderstood

Owners often respond to stagnation by:

- switching vendors
- changing tactics
- adding tools

But fragmentation rarely looks broken in isolation. Each part may appear "fine." The failure is relational, not technical.

The system doesn't average signals.

It resolves uncertainty by pulling back.

This matters most when...

Every vendor report sounds reasonable, but the business feels harder to explain than it should.

3. AI as a Pattern Amplifier

AI does not invent new rules.

It accelerates existing ones.

AI systems summarize patterns across surfaces. They reward repetition, corroboration, and stability—and surface contradiction faster than humans ever could.

AI doesn't decide what matters.

It decides what is *consistent enough* to repeat.

What breaks when this is misunderstood

When AI is treated as a shortcut, owners often:

- scale content before structure exists
- amplify vague messaging
- automate inconsistency

The result isn't penalty.

It's invisibility.

AI doesn't argue with confusion.

It avoids it.

This matters most when...

You feel pressure to "use AI" but can't clearly explain what it should reinforce.

4. Maps as Real-World Validation

Maps is not an SEO feature.

It is a reality check.

Maps helps platforms answer:

> "Is this business real, local, and behaving like legitimate businesses in this category tend to behave?"

Because dental practices are trust-sensitive, Maps quietly influences everything else—even organic and paid performance.

What breaks when this is misunderstood

When Maps is treated as just another ranking game:

- categories drift
- services blur
- proximity and legitimacy signals weaken

The system doesn't correct this for you.

It simply becomes less confident.

This matters most when...

You rank well organically but still struggle with local trust or conversion quality.

5. Ads as Stress Tests, Not Fixes

Paid ads accelerate demand.

They do not create trust.

Ads are most effective when they point toward something that already makes sense. When they don't, they surface problems faster than organic ever will.

What breaks when this is misunderstood

When ads are expected to compensate for structural gaps:

- costs rise

- conversion quality drops
- frustration increases

Ads didn't fail.

They revealed misalignment.

This matters most when...

Ads "used to work," but now feel unpredictable or expensive.

6. Compliance as a Visibility Constraint

In dental practice marketing, compliance is not just legal hygiene.

It is a risk signal.

Careful language, realistic framing, and consistency across claims reduce platform hesitation. Exaggeration, vagueness, or contradiction increase it.

Restraint scales better than boldness in trust-sensitive categories.

What breaks when this is misunderstood

When compliance is treated as an afterthought:

- messaging drifts
- disclaimers feel bolted on
- AI systems flag inconsistency

The result is not punishment—it's caution.

This matters most when...

You feel torn between sounding compelling and sounding responsible.

7. Measurement as Direction, Not Proof

Attribution simplifies reality.

It does not explain it.

Good measurement informs sequencing and prioritization. Bad measurement is used to defend decisions already made.

What breaks when this is misunderstood

When owners demand certainty from incomplete data:

- long-term investments are cut too early
- ads are over-credited
- trust-building is undervalued

Clarity doesn't come from perfect data.

It comes from asking better questions.

This matters most when...

Dashboards look busy, but decisions feel harder instead of easier.

APPENDIX B

A System Map of Modern Dental Practice Marketing

All modern dental practice marketing systems resolve to one outcome:

selection confidence.

Maps validates real-world presence.

Websites establish structural clarity.

Content demonstrates expertise over time.

Social media confirms reality and consistency.

Ads accelerate demand and test alignment.

Operations and patient experience feed back into trust signals.

AI sits above this system—not replacing it, but amplifying whatever patterns already exist.

When these elements reinforce each other, growth compounds quietly.

When they contradict each other, progress stalls without obvious failure.

This map matters because it reframes marketing from activity into architecture.

APPENDIX C

Resources for Understanding the System (Not Chasing Tactics)

This book avoids tactics intentionally. Still, many readers want to understand how these ideas connect to primary sources.

The resources below are included not for instruction, but for context—useful for understanding incentives, constraints, and system behavior.

They are best read slowly, not implemented quickly.

For those readers, we've created a small set of optional tools and decision-support frameworks at https://patientflow.us/resources-dental*, including system reviews, website clarity ratings, content strategy builders, competitor alignment reports, and vendor interview preparation.*

APPENDIX D

How to Interview Digital Marketing Companies

Most owners don't hire the wrong agency because they failed to ask the right technical questions.

They hire the wrong agency because confidence sounds like competence when pressure is high.

The questions below are not traps. They are lenses.

Why smart owners still hire the wrong partners

When marketing feels unstable, owners naturally look for certainty. Vendors who promise clarity, speed, or control can feel reassuring—even when their explanations are incomplete.

Misalignment rarely announces itself as incompetence.

It announces itself as oversimplification.

How They Think About Systems

- "How do Maps, organic search, ads, and content reinforce each other?"
- "Where do you expect friction to show up first if something isn't aligned?"

Good answers describe interaction and tradeoffs.

Weak answers isolate channels.

How They Think About AI

- "How has AI changed what you prioritize for clients like us?"
- "What mistakes are you seeing businesses make with AI-generated content?"

Good answers emphasize structure and restraint.

Weak answers emphasize speed and volume.

How They Think About Measurement

- "Which metrics do you trust least, and why?"
- "What should improve before rankings or leads do?"

Good answers acknowledge ambiguity.

Weak answers promise certainty.

How They Think About Risk

- "How do you think about compliance in dental practice marketing?"
- "What would cause you to slow down instead of pushing harder?"

Good answers treat caution as strategy.

Weak answers dismiss risk.

How They Frame Timelines

- "What should feel different in the first 90 days, even if results don't change yet?"

Good answers describe signal accumulation.

Weak answers promise visible outcomes without context.

You are not evaluating confidence.

You are evaluating clarity.

A Note on Optional Tools and Frameworks

Some owners prefer to apply these ideas using structured frameworks rather than starting from scratch.

For those readers, we've built a small set of optional decision-support tools at **Patient Flow**. They're designed to help clarify alignment, competition, messaging, and vendor decisions—without turning strategy into automation.

They are educational, optional, and meant to support judgment—not replace it.

You can find a list of recommended AI tools here: https://patientflow.us/resources-dental

A Quiet Test of Alignment

Before closing the book, consider this:

If you had to explain your marketing system in one uninterrupted sentence—without naming channels—could you?

If not, that's not a failure.

It's a signal.

Clarity usually precedes progress.

This appendix exists so you have somewhere to return when you're ready to look for it again.

Made in United States
Troutdale, OR
11/03/2026

56845370R00137